Environmental Protection and the Social Responsibility of Firms

Perspectives from Law, Economics, and Business

Bruce L. Hay
Robert N. Stavins
Richard H. K. Vietor
Editors

Resources for the Future
Washington, DC, USA

Printed in the United States of America

An RFF Press book
Published by Resources for the Future
1616 P Street NW
Washington, DC 20036–1400
USA
www.rffpress.org

Library of Congress Cataloging-in-Publication Data

Environmental protection and the social responsibility of firms : perspectives from law, economics, and business / Bruce L. Hay, Robert N. Stavins, and Richard H. K. Vietor, editors.
p. cm.
Includes index.
ISBN 1-933115-02-5 (hardcover : alk. paper) — ISBN 1-933115-03-3 (pbk. : alk. paper)
1. Industrial management—Environmental aspects. 2. Environmental protection. 3. Social responsibility of business. I. Hay, Bruce L. II. Stavins, R. N. (Robert N.), 1948- III. Vietor, Richard H. K., 1945- IV. Series.

HD30.255.E5873 2004
363.7--dc22 2004022279

The paper in this book meets the guidelines for permanence and durability of the Committee on Production Guidelines for Book Longevity of the Council on Library Resources. This book was typeset by Peter Lindeman. It was copyedited by Joyce Bond. The cover was designed by Maggie Powell.

ISBN 1-933115-02-5 (cloth) ISBN 1-933115-03-3 (paper)

About Resources for the Future *and* RFF Press

Resources for the Future (RFF) improves environmental and natural resource policymaking worldwide through independent social science research of the highest caliber. Founded in 1952, RFF pioneered the application of economics as a tool for developing more effective policy about the use and conservation of natural resources. Its scholars continue to employ social science methods to analyze critical issues concerning pollution control, energy policy, land and water use, hazardous waste, climate change, biodiversity, and the environmental challenges of developing countries.

RFF Press supports the mission of RFF by publishing book-length works that present a broad range of approaches to the study of natural resources and the environment. Its authors and editors include RFF staff, researchers from the larger academic and policy communities, and journalists. Audiences for publications by RFF Press include all of the participants in the policymaking process—scholars, the media, advocacy groups, NGOs, professionals in business and government, and the public.

We dedicate this volume to Andrew Savitz, whose enthusiasm upon his return from the Johannesburg Earth Summit inspired us to work together to explore the ideas that are at the heart of this book.

Contents

Part II: The Economic Perspective

Part III: The Business Perspective

About the Contributors

Dennis J. Aigner is professor of management and economics and Bren Fellow in Business Management at the University of California, Irvine, and Dean of the Bren School of Environmental Science and Management at the University of California, Santa Barbara. He has written widely on econometrics, statistics, microeconomics, and operations research. His current research interest is the empirical relationship between corporate environmental and financial performance.

John J. Donohue, a professor at Yale Law School, is a fellow of the National Bureau of Economic Research and writes extensively on public policy issues. He teaches courses on corporations, contracts, torts, criminal law, employment discrimination, law and economics, and law and statistics. He is the author of *Foundations of Employment Discrimination Law.*

Einer R. Elhauge is professor of law at Harvard Law School, where he teaches courses on antitrust, contracts, corporations, health care, and public choice theory. Before coming to Harvard, he was a professor of law at the University of California at Berkeley and clerked for Judge Norris on the Ninth Circuit Court of Appeals and Justice Brennan on the Supreme Court.

Daniel C. Esty is professor of environmental law and policy at Yale University, with appointments in Yale's Environment and Law schools. He is director of the Yale Center for Environmental Law and Policy and editor or author of eight books. He previously served as a senior fellow at the Institute for International Economics, a senior official in the U.S. Environmental Protection Agency, and a practicing attorney.

Bruce L. Hay is professor of law at Harvard Law School and chair of the Harvard University Advisory Committee on Shareholder Responsibility. His research interests include economic analysis of law, evidence, legal ethics, and legal procedure. His recent work includes "Manufacturer Liability for Harms Caused by Consumers to Others" (with Kathryn Spier).

Eric W. Orts is the Guardsmark Professor at the Wharton School of the University of Pennsylvania, where he is a professor in the Legal Studies and Management departments and directs the Environmental Management Program. His recent research in corporate governance and environmental management includes *Environmental Contracts: Comparative Approaches to Regulatory Innovation in the United States and Europe* (with Kurt Deketelaere).

Paul R. Portney is president and a senior fellow of Resources for the Future. His research interests include the application of cost–benefit analysis to environmental regulations. He is the coeditor (with Robert Stavins) of *Public Policies for Environmental Protection, Second Edition.*

Forest L. Reinhardt is the John D. Black Professor of Business Administration at Harvard Business School. He is the author of *Down to Earth: Applying Business Principles to Environmental Management.* His writing on business and the environment has recently appeared in *Harvard Business Review* and in the World Economic Forum's Global Competitiveness Report.

Mark J. Roe is the David Berg Professor of Corporate Law at Harvard Law School, where he teaches bankruptcy and corporate law. His recent publications include *Political Determinants of Corporate Governance* and "Delaware's Competition," 117 *Harvard Law Review* 588 (2003).

Robert N. Stavins is the Albert Pratt Professor of Business and Government at the John F. Kennedy School of Government and director of the Environmental Economics Program at Harvard University. He is a university fellow of Resources for the Future and former chairman of the U.S. Environmental Protection Agency's Environmental Economics Advisory Committee. His recent books include *Economics of the Environment, The Political Economy of Environmental Regulation,* and (with Paul Portney) *Public Policies for Environmental Protection, Second Edition.*

Richard H. K. Vietor is the Senator John Heinz Professor of Environmental Management at the Harvard Graduate School of Business Administration. He teaches courses in political economy and is author of numerous books, articles, and more than 60 cases. Among his recent books are *Contrived Competition, Business Management and the Natural Environment,* and *Globalization and Growth.*

David J. Vogel is a professor at the Haas School of Business and the Department of Political Science at the University of California, Berkeley. He has written extensively on the national, comparative, and international dimensions of consumer and environmental regulation. His recent books include *Trading Up, Benefits and Barriers,* and *The Dynamics of Regulatory Change.*

Overview

The Four Questions of Corporate Social Responsibility

May They, Can They, Should They, Do They?

Bruce L. Hay, Robert N. Stavins, and Richard H. K. Vietor

The time is ripe for critical and informed reflection on the concept of "corporate social responsibility" (CSR) in the realm of environmental protection. Such reflection can help lay the foundations for better thinking and more sensible action by providing new grounding for what has been most reasonable, by weeding out less judicious courses of action, and overall, by providing sharper focus.

At issue is the appropriate role of business with regard to environmental protection. Everyone agrees that firms should obey the law. But beyond the law—beyond complete compliance with environmental regulations—do firms have additional moral or social responsibilities to (voluntarily) commit resources to environmental protection? How should we think about the notion of firms sacrificing profits in the social interest? *May* they do so within the scope of their fiduciary responsibilities to their shareholders? *Can* they do so on a sustainable basis, or will the forces of a competitive marketplace render such efforts and their impacts transient at best? Further, *should* firms carry out such profit-sacrificing activities, despite their positive environmental impacts? Is this an efficient use of social resources? And finally, *do* firms, in fact, frequently or at least sometimes behave this way? Do some firms reduce their earnings by voluntarily engaging in environmental stewardship?

To explore these questions, Harvard Business School, Harvard Law School, and Harvard's John F. Kennedy School of Government formed a partnership. This book is the initial product of the collaboration. With support from the Provost of the University, the Harvard University Center for the Environment, the Harvard Business School, the Surdna Foundation, and the Savitz Family Fund for Environmental and Natural Resource Policy, the partnership organized a workshop on Environmental Protection and the Social Responsibility of Firms, featuring a one-day meeting of leading scholars from across the United States on December 6, 2003.

Our purpose in this book is not to serve as a cheerleading platform either for or against the corporate social responsibility movement, but to foster a careful and impartial exploration of these four questions of corporate social responsibility in relation to the environment. Up until now, public discussion has generated far more heat than light on both the normative and positive questions surrounding CSR in the environmental realm. Our objectives are to step back from the debates among activists, examine the assumptions often uncritically accepted in the debates, and provide a rigorous foundation for future research and policymaking. Claims and counterclaims regarding the social responsibility of firms in the environmental realm not only contradict each other, but also too often rest on undefined terms or confused analysis. Moreover, theoretical arguments frequently have failed to take account of whether there is relevant, supporting empirical evidence. By reflecting on what is known and taking explicit note of what is not, we hope to help build a framework for future analysis and action.

The workshop was deliberately kept small, bringing together 20 scholars for a day of discussion of the central issues, drawing on perspectives from law, economics, and business scholarship. The table on page 3 lists participants and their affiliations.

For the workshop and for this book, the partnership commissioned three papers to survey, synthesize, and extend the existing literature: one on the legal aspects of corporate social responsibility; one on the economic rationale and implications of CSR; and one on related business practice. Each paper received two formal commentaries plus extended discussion by all participants. In this book, we assemble the three papers, the six commentaries, and summaries of the respective discussions.[1] The papers and commentaries that were presented and commented upon at the conference have been further reviewed and edited, with authors having the opportunity to update their contributions and refine and clarify their arguments.

The Legal Perspective

Our point of departure involves the law. Does management have a fiduciary duty to maximize corporate profits in the interest of shareholders, or can it sacrifice profits by voluntarily exceeding the requirements of environmental law? Einer Elhauge, a professor at the Harvard Law School, wrote the volume's first paper, "Corporate Managers' Operational Discretion to Sacrifice Corporate Profits in the Public Interest," to address this and related problems.

1. Initial drafts of the discussion summaries were prepared by Darby Jack, a PhD student in Public Policy at Harvard, and Sam Walsh, a student at Harvard Law School and the John F. Kennedy School of Government.

TABLE 1. *Participants at December 2003 Meeting on Law, Economics, and Business Approaches to Corporate Social Responsibility*

Dennis Aigner	Bren School of Environment, University of California, Santa Barbara
Max Bazerman	Harvard Business School
John Coates	Harvard Law School
Cary Coglianese	John F. Kennedy School of Government, Harvard University
John Donohue	Stanford Law School
Einer Elhauge	Harvard Law School
Daniel Esty	Yale School of Forestry and Environmental Studies
Bruce Hay	Harvard Law School
William Hogan	John F. Kennedy School of Government, Harvard University
Paul Kleindorfer	The Wharton School, University of Pennsylvania
Charles Kolstad	Department of Economics, University of California, Santa Barbara
James Krier	University of Michigan School of Law
Thomas Lyon	School of Business, Indiana University
Eric Orts	The Wharton School, University of Pennsylvania
Paul Portney	Resources for the Future
Forest Reinhardt	Harvard Business School
Mark Roe	Harvard Law School
Robert Stavins	John F. Kennedy School of Government, Harvard University
Richard Vietor	Harvard Business School
David Vogel	Haas School of Business, University of California, Berkeley

Note: Five others who were scheduled to participate could not attend because of severe weather conditions: Mark Cohen, Owen Graduate School of Management, Vanderbilt University; Stuart Hart, Kenan-Flagler Business School, University of North Carolina; Kathy Segerson, Department of Economics, University of Connecticut; Richard Stewart, School of Law, New York University; and Lawrence Summers, Harvard University.

As a practical matter, firms' managers may generally may undertake environmental initiatives beyond what is required by regulation without running into legal trouble. Managers' legal duties to shareholders are governed by the "business judgment rule," which gives them almost unlimited discretion to use corporate resources as they see fit (provided they are not lining their own pockets, which will trigger greater legal scrutiny). According to the business judgment rule, any measure that even arguably may increase profits—for example, by improving public relations, increasing consumer loyalty, or boosting employee morale—will not be second-guessed by courts. This is why firms can sponsor sporting events or cultural activities, pay above-market wages, refuse to do business in countries with repressive governments, and take other steps that may seem to the observer to sacrifice profits. The traditional rationale for the business judgment rule is that it is often difficult to distinguish profitable from unprofitable ventures, and that courts are less qualified than managers to make such judgments. The rule quite clearly extends to environmental measures: if a company's managers decide, for example, to use "green" inputs, to devise cleaner production technologies, or to dispose of their waste more safely, no court will stop them from doing so, no matter how disgruntled the firms' shareholders may be at such acts of public charity. For all a judge knows, such measures—particularly when they are well publicized—will add to the firm's bottom line in the long run by increasing public goodwill.

But these generalizations about the law leave unanswered some important questions about environmental CSR measures. First, will courts really permit managers to expend corporate resources on any public-spirited action they choose, so long as their lawyers can articulate some theory, however far-fetched, that the action is in the company's interests? Thus far, courts have not stood in the way of voluntary environmental initiatives by firms. But it remains to be seen how far courts will allow firms to go in this direction: perhaps some voluntary environmental measures may be so flagrantly unprofitable from shareholders' perspectives that they will not survive scrutiny even under the business judgment rule. Second, if courts are—or turn out to be—so lenient that they will permit any and all environmental CSR measures that promote the public good, should the business judgment rule be changed to make courts less solicitous of such measures? Should courts—or legislatures—adopt a rule that would enable shareholder lawsuits to block environmental measures that demonstrably harm shareholders for the benefit of the public?

Elhauge challenges the conventional wisdom of the law and economics literature, which generally claims that managers have (and should have) a legal duty to maximize corporate profits. He argues that, as a positive matter, managers have broad discretion to diverge from the goal of profit maximization. The business judgment rule is one source of such discretion; and Elhauge argues that managers may receive protection from this rule even without articulating a profit-maximizing rationale for their actions. Another source of managerial discretion, he notes, is the decline of the market for corporate takeovers; a vigorous market for takeovers would punish managers who sacrificed profits, and so would keep them focused on the bottom line. But the market for takeovers was largely gutted in the 1980s and 1990s by courts and legislatures, which have permitted managers to use "poison pills" and other antitakeover strategies.

Elhauge also argues that, as a normative matter, managers *should* have the discretion they enjoy to sacrifice profits for the sake of enhancing social welfare. He maintains that external enforcement of social welfare policies—for example, government regulation—is inefficient on a firm-by-firm basis. Thus if corporate managers make decisions to fill gaps—as in the case of an "underregulated" firm that decides to go further than the regulations require—then overall social welfare may be enhanced. Elhauge extends his arguments beyond environmental performance, including a comparison of operational decisions (over which managers presumably have expertise) to charitable donations (to which corporate expertise does not extend).

Mark Roe and John Donohue, of Harvard and Yale Law Schools, respectively, provide commentaries on Elhauge's paper. Both agree with aspects of his analysis but part ways with him in important respects. Roe raises doubts about Elhauge's positive claims about the law. He believes that courts will not permit managers to avowedly sacrifice profits for the sake of social welfare. He also contends that the market for corporate control may be vigorous enough to keep managers from substantially sacrificing profits for social purposes. In addition, Roe argues that even if the legal environment permits managers to sacrifice profits, significant nonlegal pressures are exerted for them to focus on the bottom line. Among such nonlegal forces, he cites heightened product-market competition, rapid technology change, and social expectations.

Donohue joins Roe in suggesting that Elhauge may have overstated the degree to which the law permits managers to diverge from the goal of profit maximization. He agrees that managers can indeed sacrifice profits for a variety of humanitarian or philanthropic purposes but thinks they must do so quietly and must tailor their rationale "to promote the long-term interest of the company." Donohue also doubts whether managers should be given, as a matter of policy, the discretion that Elhauge advocates. He suggests that managers often lack the expertise to determine what advances overall social welfare and should confine their decisions to areas in which they have a comparative informational advantage.

In the discussion that follows, several participants express considerable surprise: economists and business academics know little about the business judgment rule and the corresponding degree of latitude that the law gives managers to make decisions that may or may not raise the firm's bottom line. Once that hurdle is overcome, discussion turns to the normative question of social efficiency: how can managerial discretion make economic trade-offs efficiently? Lawyers respond that managers are better informed about efficient trade-offs than regulators, and that regulation is not as efficient across diverse industry structures as is firm-specific, managerial knowledge. Finally, discussion turns to increasing constraints on corporate latitude, from global competition to scientific complexity, possibly mooting the issues of managerial flexibility.

The Economic Perspective

Two economic questions—one positive, one normative—are the central focus of the second paper in this volume, "Corporate Social Responsibility: An Economic and Public Policy Perspective," by Paul R. Portney of Resources for the Future.

First, *can* firms carry out beyond-compliance environmental protection efforts on a sustainable basis, or will the forces of a competitive marketplace essentially preclude such actions, at least in the long term? It would seem that for firms that enjoy monopoly positions or produce products for well-defined niche markets, such extra costs could well be passed on to customers. On the other hand, it would seem that firms that exist in competitive industries—particularly firms that produce commodities—would find it difficult or impossible to pass on such voluntarily incurred costs to customers, and would therefore have to absorb those extra costs in the form of reduced profits, reduced shareholder dividends, or reduced compensation, suggesting that, in the face of competition, this would hardly be sustainable. Second, if firms may carry out such profit-sacrificing activities (that is, are allowed to, by law), and can do so, *should* they, despite the presumed positive environmental impacts? Is this likely to lead to an efficient use of social resources?

Portney begins with a definition: "CSR [is] a consistent pattern, at the very least, of private firms doing more than they are required to do under applicable laws and regulations governing the environment, worker safety and health, and investments in the communities in which they operate." Using this as his starting point, Portney reviews the conceptual arguments for government regulation, and asks whether it is reasonable to anticipate that governments systematically underregulate in this area.

Portney turns to CSR, offering two fundamental reasons why firms might engage in the practice: because it is a moral obligation or because it is in a firm's economic interest to do so. Focusing on the second, economic justification, Portney suggests that CSR behavior can be rewarded because (1) some consumers may reward it, such as through purchases of differentiated products; (2) firms may find it makes it easier to attract and retain skilled and highly motivated employees; (3) it may reduce firms' costs of capital; (4) it may engender favorable treatment by regulators and local communities; and (5) it may make it possible for firms to preempt more onerous regulations or even influence the form of regulation in ways that raise costs for competitors. In order to examine these potential motivations for CSR behavior, Portney reviews the extensive literature that has developed. He laments that none of the studies derive testable hypotheses from theoretical models of the firm, and he finds little empirical evidence in support of the theoretical claims.

The balance of Portney's paper focuses on the key normative question: under what conditions are firms' CSR activities likely to be welfare-enhancing? He finds that this is most likely to be the case if firms pursuing CSR strategies are doing so because it is good business—that is, profitable. On the other hand, for more costly CSR investments, concern exists about the opportunity costs that will be involved for firms. Furthermore, in the case of firms that behave strategically with CSR to anticipate and shape future regulations, welfare may be reduced if the result is less stringent standards (that would have been justified). Likewise, international harmonization, "although well-intentioned, ... overlooks the fact that differences in incomes across countries and over time give rise to very different demands for (and hence valuations of) environmental protection, workplace safety, and other objects of CSR." Portney concludes that welfare almost surely will be reduced by such harmonization. In concrete terms, this means that in developing countries, society may benefit more from lower product prices or higher wages than from environmental quality measures.

Finally, Portney laments that discussions of CSR trivialize what are the most important social contributions of firms: the production of goods and services people value, and the provision of jobs. In addition, he questions why beyond-compliance CSR that is profitable should be thought of differently than any other profit-maximizing activities. And he worries that activists may push for CSR because they do not trust the public sector to provide the pubic goods in question, which may simply be evidence of the fact that the social benefits do not justify the social costs.

Commentaries on this paper are provided by Daniel Esty of Yale University and Dennis Aigner of the University of California at Santa Barbara. Esty concurs with Portney that CSR often may be well-intentioned but wasteful, in that it does not enhance overall social welfare. He is skeptical of the social gain from CSR. "Do we really want corporations guided by something other than the law?" asks Esty. What if they get it wrong? Esty goes on to offer five ways to reconceptualize CSR: as a check on "regulatory failure"; a mechanism for regulatory flexibility; a procedural guide to corporate interaction with stakeholders; a way to hold corporations "accountable" for their impacts on society; or a "winner's tax." After exploring these, Esty concludes that rigorous definition of CSR has only begun, and that more serious work is needed if it is to be a meaningful guide to business.

In his commentary, Dennis Aigner congratulates Portney for his fair treatment of CSR but challenges his logic on several grounds. Aigner considers Portney to be excessively harsh on the empirical research to date for not having examined testable hypotheses. He acknowledges that the data, going back to the 1970s, and the econometric analyses are weak—"lousy," in fact. But he suggests that better, more recent data are available, indicating links between corporate environmental and financial performance. Aigner also takes issue with Portney's warning that CSR activities can do more harm than good, and he disagrees with Portney's view that discussions about CSR typically underplay the truly valuable things that corporations do—that is, create products that consumers value and provide jobs.

The general discussion is dominated by two themes: whether or not Portney's definition of CSR is adequate, and the wisdom of incorporating in the definition the idea that corporations need to sacrifice profits. Some suggest that "beyond compliance" is a binary variable: that the law is often rough, open to wide judicial interpretation. Although many discussants seem to believe that CSR should merely entail "beyond compliance," others insist on limiting the scope of real CSR to the notion of "profit-sacrificing behavior." In response, others argue that measurement of such behavior is exceedingly difficult, even more so for smaller corporations and privately held companies.

The Business Perspective

The final paper in this volume, by Forest L. Reinhardt of the Harvard Business School, is "Environmental Protection and the Social Responsibility of Firms: Perspectives from the Business Literature." In this essay, Professor Reinhardt surveys the literature on the environmental performance of firms. Like Portney, Reinhardt starts with definitions, including a definition of "sustainable development." He contrasts this with definitions of corporate social responsibility (CSR) and David Baron's notion of corporate social performance (CSP). Sustainable development, writes Reinhardt, entails maintaining (or increasing) a nation's stock of assets over time—both natural and man-made—regardless of motivation. The same is true, concludes Reinhardt, for corporate social responsibility; that is, it entails a redistribution from firm to populace, but not necessarily with altruism as a motive.

The logic of this leads Reinhardt to argue that there is no need for CSR "if it always pays to be green." All firms would be green (that is, exceed compliance with environmental regulations), just to serve their shareholders. From this it follows that it rarely pays to be green, and that "relatively few $20 bills can be found lying on the pavement." But situations exist in which it does indeed pay. Where one can increase customers' willingness to pay, reduce one's costs, manage future risk, or anticipate/defer costly governmental regulation, then it may pay to be green. Here Reinhardt is echoed by Esty and Aigner, who worry that some CSR measures may be cynical ploys to ward off socially desirable regulation. Reinhardt acknowledges the existence of real opportunities for firms (such as Patagonia and DuPont) but does not support broad claims of "opportunities of potentially staggering proportions."

Reinhardt considers the statistical evidence, noting problems of isolating the effects of environmental action on overall profitability, problems of data selection (large and successful firms), and problems of identifying causality. He makes a plea for empirical case studies to tie economics to organizational theory. In his conclusions, Reinhardt draws implications of the literature for activists, regulators, investors, managers, and business scholars.

In his commentary, David Vogel of the University of California at Berkeley generally agrees with Reinhardt, while taking issue with several points. He starts by noting that Europeans are more willing than Americans to pay for environmental amenities. Ecomarketing seems to work better there. Likewise, on cost reduction, Vogel is somewhat more optimistic. Managers, concludes Vogel, are invariably myopic. Until they get focused on a set of issues outside of their managerial conventions, they can easily miss opportunities.

Vogel goes on to criticize optimists such as Stuart Hart, who see huge opportunities for corporate environmentalism. Scholars of this persuasion disregard public policy, writes Vogel. They do not see the immense environmental gains attributable to regulation, nor the fact that most environmental regulation has been opposed by the businesses to be regulated. If the world is to be more "sustainable," argues Vogel, then business lobbies should support environmental regulation, not oppose it.

Eric W. Orts, of the Wharton School at the University of Pennsylvania, takes a different approach in his commentary on the Reinhardt essay. After noting that few "win–win" solutions probably exist, he argues for proactive environmental management as a self-imposed, moral constraint. In other words, he believes there is a compelling need for an ethical response to severe environmental risks imposed on business in contemporary society. Picking up on Baron's distinction between CSP and CSR, Orts objects to "strategic CSR"—the idea of doing something green as part of a profit-maximizing strategy and labeling it CSR. Rather, for Orts, the correct definition of CSR enters "only when the profit motive runs out." Also, whereas Reinhardt views risk management as affecting the expected value of firms, Orts argues that because systemic environmental risks do not affect cash flows, they can be ignored by diversified investors. Orts writes that many firms apply global safety standards on moral grounds, rather than as risk management.

Orts implicitly equates Reinhardt with Milton Friedman, who argues against CSR and contends that business should trust in public policy. According to Orts, this is naive. To the contrary, Orts maintains that business should guard against an overly self-interested view of (environmental) legislation, instead taking a broad, public-interest perspective. Overall, Orts makes a plea for ethical constraints and duties of business to extend to environmental considerations. He indicates that CSR entails an area of ethical constraints, in which costs and profitability are scarcely relevant.

Intense discussion of these topics follows. The question of how to think about ethics in developing countries is raised. Some discussants suggest that firms themselves need to assess risk, as poor workers may not understand trade-offs. Others claim that governments are responsible for just such policies. From there, discussion returns to the role of CSR in environmental policy. One discussant claims that public policies in the environmental realm typically emerge in response to crises or with the support of the private sector. Others object, claiming reliance on a different history of environmental policies. In response, another argument is raised: that the scale of global firms and environmental problems may render this history moot. In the end, discussants remain divided as to whether volunteerism is significant, much less appropriate.

The three central chapters presented in this book, along with the six related commentaries and three respective discussions, constitute a beginning for careful exploration of corporate social responsibility in the environmental domain. May firms engage in CSR, beyond the law? An affirmative answer seems fairly clear, though not without controversy. Can firms do so on a sustainable basis? Outside of monopolies and limited niche markets, the answer to this question is considerably less clear. Should they carry out such beyond-compliance efforts, even when doing so is not profitable? Here—if the alternative is sound government policy—the answer is even less clear. And the positive question—do they generally carry out such activities?—seems to lead to a negative assessment in general, at least if we restrict our attention to real cases of "sacrificing profits in the public interest."

Acknowledgments

We gratefully acknowledge support received from the Harvard University Center for the Environment, the Provost of Harvard University, the Harvard Business School, and the Surdna Foundation.

PART I

The Legal Perspective

Einer R. Elhauge

Comments

John J. Donohue
Mark J. Roe

Corporate Managers' Operational Discretion to Sacrifice Corporate Profits in the Public Interest

Einer R. Elhauge

Let's start concrete before we get theoretical. Suppose clear-cutting is profitable and legal, but is nonetheless regarded as environmentally irresponsible under prevailing social norms. Can management of a timber corporation decline to clear-cut its timberland even though it sacrifices profits? One might be tempted to evade the question by claiming that being environmentally responsible is profitable in the long run, either because it preserves the forest for future harvesting or because it maintains a public goodwill that aids future sales. But suppose, in an incautious moment, management admits that the present value of those future profits from not clear-cutting cannot hope to match the large current profits that clear-cutting would produce.

My answer to this question will challenge the canonical law and economics view that corporate managers do and should have a duty to profit-maximize because such conduct is socially efficient given that general legal sanctions do or can redress any harm that corporate or noncorporate businesses inflict on others.[1] My contention is that this canonical view is mistaken both descriptively and normatively.

In fact, the law does give corporate managers considerable implicit and explicit discretion to sacrifice profits in the public interest. They would have it

---

1. *See, e.g.*, Robert Clark, Corporate Law 17–21, 30–32, 603, 677–694, 702 (1986); American Bar Association Committee on Corporate Laws, *Other Constituency Statutes: Potential for Confusion*, 45 Bus. L. Rev. 2253, 2257–61, 2269–70 (1990) [hereinafter "ABA"]; Milton Friedman, *The Social Responsibility of Business Is to Increase Its Profits*, The New York Times, Sept. 13, 1970 (Magazine at 33); Henry Hansmann & Reinier Kraakman, The End of History for Corporate Law, 89 Geo. L.J. 439, 440–42 (2001); Jonathan Macey, *An Economic Analysis of the Various Rationales for Making Shareholders the Exclusive Beneficiaries of Corporate Fiduciary Duties*, 21 Stetson L. Rev. 23, 23, 40–41 (1991).

even under a profit-maximization goal because minimizing total agency costs requires giving managers a business judgment rule deference that necessarily confers such profit-sacrificing discretion. The alternative of eliminating this discretion by creating a legally enforceable duty to profit-maximize would put the litigation process, rather than managers subject to market processes, in charge of operational decisions. This would surely lower shareholder profits and increase total agency costs given the length, cost, and high error rate of the litigation process. Even if courts could figure out whether the conduct failed to maximize profits in the short run, it would be too difficult to tell whether it might increase profits in the long run because of increased goodwill or similar effects. Still greater difficulties are raised by the disjunction between ex ante and ex post profit maximization created by claims that efficient implicit contracts or social understandings sometimes involve others conferring sunk benefits on a corporation expecting that managers will have discretion to reciprocate later in a decision that sacrifices profits ex post (ignoring the sunk benefits) but that maximizes profits ex ante (because necessary to induce the sunk benefits).

Nor is profit maximization socially efficient because even optimal legal sanctions are necessarily imperfect and require supplementation by social and moral sanctions to fully optimize conduct. Accordingly, pure profit maximization would worsen corporate conduct by overriding these social and moral sanctions. In addition to being socially inefficient, this would harm shareholder welfare whenever shareholders value the profits less than avoiding social and moral sanctions. Where a controlling shareholder exists, he is exposed to social and moral sanctions and has incentives to act on them, and thus is well placed to decide when to sacrifice corporate profits in the public interest. But the structure of large, publicly held corporations insulates dispersed shareholders from social and moral sanctions and creates collective-action obstacles to acting on any social or moral impulses they do feel. Thus in public corporations, optimizing corporate conduct requires giving managers some operational discretion to sacrifice profits in the public interest, even without shareholder approval, because managers are, unlike shareholders, sufficiently exposed to social and moral sanctions.

Managerial incentives to be excessively generous are constrained by product markets, capital markets, labor markets, takeover threats, shareholder voting, and managerial profit sharing or stock options. These market forces generally mean that any managerial decision to sacrifice profits in the public interest substitutes for more self-interested profit-sacrificing exercises of agency slack. Managerial discretion is further constrained by legal limits on the amount of profit sacrificing, which become much tighter when market constraints are inoperable because of last-period problems.

To avoid possible misunderstanding, let me make clear what I am *not* saying. I am not saying that managers have a legally enforceable *duty* to sacrifice corporate profits in the public interest; I am saying that they have *discretion* to do so. Nor am I denying that managers' *primary* obligation is or should be to make profits, or saying that their discretion to sacrifice profits is or should be increased or boundless. I am saying that this obligation is not and should not be *exclusive*, but that instead managers do and should have some limited discretion to temper it in order to comply with social and moral norms. Managers' existing profit-sacrificing discretion is kept desirable precisely because it is bounded by the market and legal constraints I have just outlined.

I also am not saying that there is an "objective" public interest, let alone that courts can and must identify it to determine whether managers are properly exercising their discretion. By sacrificing profits "in the public interest," I simply mean to describe cases where managers are sacrificing corporate profits in a way that confers a general benefit on others, as opposed to financial benefits on themselves and intimates, the latter of which courts do seriously police under an enforceable duty of loyalty. Of course, people disagree about which efforts to confer general benefits on others are truly desirable. Some think that clear-cutting is horrific; others that it is perfectly fine or justifiable if it increases employment. Whether a discretion to benefit others will be exercised in a way that is truly desirable depends largely on whether the social and moral sanctions that influence the exercise of that discretion move behavior closer to socially desirable outcomes or farther from them. My analysis assumes only that our social and moral sanctions have enough general accuracy that overall they move us closer to the outcomes that society deems desirable rather than being affirmatively counterproductive.

But by focusing on profit-sacrificing conduct, I do want to cut off the usual reaction of "fighting the hypothetical" with claims that the socially responsible conduct at issue really increases profits in some indirect way. I understand that broader definitions of corporate social responsibility exist, one of which includes any corporate conduct that goes beyond legal compliance, even if it is profit-maximizing. Such broader definitions arguably may be of greater practical interest to activists interested in getting corporations to engage in certain conduct. After all, it is much easier to persuade corporations to stop clear-cutting if one can show that doing so is not only good, but profitable. However, such profitable activities raise no real issue of legal or normative interest. Of course, corporate managers can and should do good when it maximizes profits: what could be the argument to the contrary? The serious question is whether they can and should do good when it decreases profits. I wonder also whether socially responsible conduct that maximizes profits is really even of much practical interest. Agitating for corporations to engage in responsible conduct that increases their profits is a lot like saying there are $20 bills lying on the sidewalk that they have missed. Maybe sometimes they have missed them, but they already have ample incentives to recognize and act on such profit-maximizing opportunities. Arguments that socially responsible conduct would increase profits thus are probably less about identifying profit-maximizing opportunities that corporations have missed than about helping create a patina of conceivable profitability that makes it easier for managers to engage in conduct that really sacrifices expected corporate profits.[2] In any event, it is implausible to think that *all* socially beneficial corporate conduct conveniently happens to be profit-maximizing, and what requires analysis is the portion that doesn't.

Finally, this article is limited to justifying the doctrine that gives corporate managers discretion to sacrifice corporate profits in the public interest when making operational decisions. I do not have space here to extend the normative analysis to the legal doctrines that give managers discretion to make donations or block takeovers, or that give managers a duty to comply with the law even when that would sacrifice corporate profits. I will show elsewhere that those doctrines also can best be justified as responses to the social and moral insulation of shareholders, and indeed that the contours of those doctrines are hard to explain otherwise.[3] But that explication will have to await another occasion, as will my analysis of whether any of these doctrines should be understood as mandatory or, rather, as default rules that corporations can alter in their charters.

The Social Regulation of Noncorporate Conduct

It helps to begin with some baseline understanding about how societies regulate noncorporate conduct. Much, but not all, of that regulation is legal. The law prohibits certain conduct and imposes sanctions on the prohibited conduct. But for well-known reasons, these legal sanctions are imperfect. Part of the reason is that our lawmaking processes are inevitably imperfect because of both interest group influence and the lack of any perfect means of aggregating preferences about what the law should be.[4]

But a more fundamental reason is that imperfect legal sanctions are in fact optimal. Even in an ideal world with perfectly unbiased decision-making processes, legal sanctions can never be made sufficiently precise to deter or condemn all undesirable activity because we lack perfect information and cannot perfectly define or adjudicate undesirable activity.[5] Trying to eliminate those imperfections in information and adjudication would be not only unfeasible and

---

2. The other argument offered for defining corporate social responsibility as conduct that exceeds legal compliance is that it is more clear and measurable. But so many behavioral choices could be said to exceed bare legal compliance that calling all of it corporate social responsibility reduces the concept to meaninglessness. For example, if a firm decides not to come close to violating the patent of another firm, it exceeds legal compliance, but is that really corporate social responsibility? To make the concept meaningful, such a definition must implicitly be limited to conduct that exceeds legal compliance *in some socially desirable way*. But such a limit just begs the question of what legal conduct is socially desirable, and deprives the definition of its supposed advantage in clarity.

3. For a full discussion that includes all the relevant issues, see Einer Elhauge, *Sacrificing Corporate Profits in the Public Interest,* 80 N.Y.U. L. Rev. (June 2005).

4. *See generally* Einer Elhauge, *Does Interest Group Theory Justify More Intrusive Judicial Review?*, 101 Yale L.J. 31, 35–44, 101–02 (1991) (summarizing literatures on interest group influence and the inevitable imperfections with any system of aggregating preferences).

5. *See* Stephen Bundy & Einer Elhauge, *Knowledge About Legal Sanctions*, 92 Mich. L. Rev. 261, 267–79 (1993); Reinier Kraakman, *Gatekeepers*, 2 J. Law Econ & Org. 53, 56–57 (1986).

costly, but also undesirable in principle, given the harms that perfect surveillance would impose.[6] Even if we could eliminate imperfect information by constantly videotaping everyone at zero financial cost, we probably would not find it worth the harm to privacy and the resulting deterrence of innovation and desirable spontaneous interaction. Nor would it eliminate uncertainties about how best to interpret the videotapes. Given such inevitably imperfect information and adjudication, the law can never perfectly distinguish between desirable and undesirable conduct, and thus the best possible sanctions can do no better than to strike the optimal balance between underdeterring undesirable conduct and overdeterring desirable conduct.[7]

This goes beyond the argument that illegal activity often goes underpunished,[8] for one implication of modern analysis of optimal legal sanctions is that the distinction between defining rules of conduct and enforcing them is not that sharp: both are inevitably imprecise because of imperfect information and enforcement. For example, rules of conduct often are defined in terms of objective or readily identifiable factors in order to render information within the control of one party legally irrelevant, even though that information would be pertinent to the desirability of the conduct.[9] More generally, the legal system frequently chooses rules over open-ended legal standards that correspond more closely to the desirability of the conduct because the latter is both more expensive to administer and more likely to be applied erroneously, given imperfect information and errors in weighing information. But because rules don't use standards that incorporate all the factors that bear on conduct desirability, rules are necessarily over- and underinclusive on their face. Indeed, even the most open-ended legal standards (such as the antitrust rule of reason) have this feature to some extent because they do not include *all* factors that might bear on the desirability of the conduct, but limit the inquiry to some defined set of factors. One could try to make legal rules very broad to eliminate any underinclusion of undesirable conduct, but that would create excessive costs in overincluding desirable conduct. Thus even the most efficient and socially optimal legal rules generally underinclude some significant degree of undesirable conduct. This is true no matter what strategy the law adopts on rules versus standards, standards of proof, and size of penalties, because all raise the problem that expected sanctions can be increased for undesirable conduct only at the cost of increasing them for desirable conduct.

This system of necessarily imperfect legal sanctions is supplemented by a system of economic sanctions.[10] Even when legal remedies would not suffice to

6. *See* Bundy & Elhauge, *supra* note 5, at 268-69.

7. *Id.* at 267–79.

8. CLARK, *supra* note 1, at 684–87.

9. The analysis in this paragraph summarizes Bundy & Elhauge, *supra* note 5, at 267–79.

10. *See* David Charny, *Nonlegal Sanctions in Commercial Relationships*, 104 HARV. L. REV. 373, 392–93 (1990); Kent Greenfield, *Ultra Vires Lives! A Stakeholder Analysis of Corporate Illegality (with Notes on How Corporate Law Could Reinforce International Law Norms)*, 87 VA. L. REV. 1279, 1336–37 (2001) (collecting sources).

deter us from engaging in certain undesirable conduct, we might hesitate from doing so because our reputations would suffer, causing others to stop doing business with us. These economic sanctions could make it profitable to forgo that undesirable conduct. For example, if we run a sole proprietorship that is considering whether to clear-cut, and we know that doing so will cause many consumers to refuse to buy from us, that would be an economic sanction.

Unfortunately, such economic sanctions are also likely to be imperfect for various reasons. Those harmed by our actions may not have a relationship with us that allows them to impose economic sanctions. Even if they do, they may not be informed enough to impose sanctions or may not be able to inflict a large enough economic sanction to deter the misconduct. When many parties are harmed, they also may have collective-action problems that mean none of them has incentives to engage in individually costly decisions to impose economic sanctions.

For example, consumers who care about the environment face information problems in determining what our business did, whether that conduct was desirable given all the facts, and what downstream products incorporate inputs from our business. The consumer buying furniture would, for example, have a hard time knowing whether that furniture uses our lumber, let alone whether what we did really constituted clear-cutting, or whether that was undesirable given the trade-offs with local employment. Further, consumers have motivational problems because generally they are not the same set of people who suffer the harm from the undesirable conduct. If our actions harm the environment in Oregon, consumers located in other states may conclude that the clear-cutting does not harm them much, leaving them insufficiently motivated to sanction it. Social or moral sanctions may motivate consumers located elsewhere to boycott clear-cut lumber, thus generating an economic sanction. But that is unlikely because, like shareholders, consumers are to a large extent insulated from such social and moral sanctions. In fact, empirical evidence indicates that economic sanctions for crimes (such as environmental crimes) that harm unrelated third parties are far lower than those that harm suppliers, employees, or customers as buyers.[11]

Finally, even if perfectly informed and motivated, consumers will face collective-action problems. Each consumer will know that her individual purchase decision will determine whether she gets the best-priced good, but that the loss of one sale will have little effect on a businesswide decision about whether to engage in antisocial conduct. For example, suppose our furniture sells at a $1 discount, given the lower costs of clear-cutting, but imposes a social harm of $10 per piece of furniture that consumers understand and care about. Each individual consumer has incentives to buy our furniture to get the $1 discount regard-

11. *See* Cindy R. Alexander, *On the Nature of the Reputational Penalty for Corporate Crime: Evidence,* 42 J.L. & Econ. 489, 490–91, 504, 522–23 (1999); Jonathan M. Karpoff & John R. Lott, Jr., *The Reputational Penalty Firms Bear from Committing Criminal Fraud,* 36 J.L. & Econ. 757(1993).

less of what she assumes the other consumers will do. If she assumes the other consumers are not going to stop buying because of the clear-cutting, then she knows declining to buy our furniture won't stop the $10 harm from occurring but will cost her $1. If she assumes the other consumers are going to stop buying because of the clear-cutting, then she knows that the $10 harm will be stopped regardless of what she does, so she might as well take the $1 discount. Such collective-action problems generally make economic sanctions ineffective when they require numerous consumers to take action against their economic interest. Consistent with this, substantial economic sanctions are typically imposed in cases where the customers are government agencies that lack such collective-action problems, rather than private parties who have them.[12]

In short, like legal sanctions, economic sanctions are important but inevitably imperfect. We thus cannot assume that enlightened self-interest will suffice to optimize behavior. Instead, optimizing conduct requires supplementing legal and economic sanctions with a regime of social and moral sanctions that encourages each of us to consider the effects of our conduct on others, even when doing so does not increase our profits.[13] Social sanctions might include active negatives such as the embarrassment of bad publicity, the reproach of family and friends, the pain of enduring insults and protests, or being disdained or shunned by acquaintances and strangers. It might also include simply losing the pleasure that comes with knowing others think well of us. People can be strongly motivated by the desire to gain social prestige, respect, or esteem, which in turn others can withhold passively at no or little cost to themselves. All these social sanctions can injure and deter us even if they don't cost us any money.[14] Moral sanctions include the guilt or self-loathing we experience for violating moral norms; the loss of pleasurable feelings of virtue, inner peace, or satisfaction; and the effect of any moral norms that might make certain choices unthinkable regardless of how much they might benefit us.[15]

---

12. *See* Alexander, *supra* note 11, at 491–92, 494–95, 505, 523.

13. For some excellent works that review the voluminous literature, *see* Richard H. McAdams, *The Origin, Development, and Regulation of Norms,* 96 MICH. L. REV. 338, 339–54 (1997); Eric A. Posner, *Efficient Norms*, in 2 THE NEW PALGRAVE DICTIONARY OF ECONOMICS AND THE LAW 19–23 (Peter Newman ed., 1998); STEVEN SHAVELL, FOUNDATIONS OF ECONOMIC ANALYSIS OF LAW 598–646 (2004); Cass R. Sunstein, *Social Norms and Social Roles,* 96 COLUM. L. REV. 903, 904–47.

14. My terminology thus differs from that of others who use the terms "social norms" or "social sanctions" to also cover what I would call "economic sanctions" because they affect whether the conduct is profitable. *See, e.g,* ERIC A. POSNER, LAW AND SOCIAL NORMS 5, 7–8, 15–21 (2000); Robert D. Cooter, *Three Effects of Social Norms on Law: Expression, Deterrence, and Internalization,* 79 OR. L. REV. 1, 5 (2000).

15. *See* SHAVELL, *supra* note 13, at 600–01. A much-debated issue is whether moral norms operate as internal costs that influence individual cost–benefit choices or as a sort of moral reasoning that precludes cost–benefit trade-offs entirely. *See* Kaushik Basu, *Social Norms and the Law,* 3 NEW PALGRAVE ENCYCLOPEDIA OF LAW AND ECONOMICS 476, 477 (1998) (modeling as limiting the feasible choice set); Robert Cooter, *Expressive Law and Economics,* 27 J. LEGAL STUD. 585 (1998) (modeling as a cost); Martha C. Nussbaum, *Flawed Foundations: The Philo-*

The social efficiency of a social or moral norm does not mean that compliance with it is individually profitable and that thus social and moral sanctions are unnecessary. Often, social and moral sanctions are efficient precisely because they can induce each of us to engage in conduct that collectively is beneficial for us all despite the fact that it is individually unprofitable for anyone who engages in it.[16] Other times, social or moral sanctions are efficient because they enforce informal understandings or norms of trust that are more efficient than explicit contracting and require ex ante commitments to behave in ways that will be unprofitable ex post.[17] The norm of tipping for good waiter service is an example.[18] So is the norm of complying with legally unenforceable promises even when it has become inconvenient.[19] In businesses, the typical example involves others, such as workers or suppliers, making firm-specific investments that increase the business's efficiency because they trust that the business will comply with social or moral norms against opportunistically exploiting those investments later by failing to reward them.[20] Such a norm is efficient ex ante, but compliance with it after sunk benefits are received can be ex post unprofitable and thus require nonmonetary social or moral sanctions for enforcement.

Such social and moral sanctions are important, perhaps even more important than legal and economic sanctions. Consider your own behavior. To what extent is your day-to-day behavior really altered by legal and economic sanctions, rather than by social and moral norms? For most of us, I expect the answer is mainly by the latter. We comply with social promises and hold doors for strangers, and we refrain from lying or abusing each other's trust, even when doing otherwise is legal and personally beneficial, and this is desirable because others reciprocate by following the same norms in ways that benefit us even more. We wouldn't steal or commit murder even if those weren't crimes. Indeed, the degree of legal compliance in society cannot be explained without social or moral sanctions, given that legal and economic sanctions frequently are insuffi-

---

sophical Critique of (a Particular Type of) Economics, 64 U. CHI. L. REV. 1197, 1211 (1997) (criticizing modeling as a cost); SHAVELL, *supra* note 13, at 604 (modeling as a cost). For my purposes, the difference does not matter because either would have the same implications for my analysis. Given that one can think of the latter view as just a special case where moral sanctions are infinite, for convenience I will use the term "moral sanction" to encompass both views.

16. *See* Robert D. Cooter, *Decentralized Law for a Complex Economy: The New Structural Approach to Adjudicating the New Law Merchant,* 144 U. PENN. L. REV. 1643, 1657–77 (1996); McAdams, *supra* note 13, at 344 & n.25; BOB ELLICKSON, ORDER WITHOUT LAW 123–26, 167–83 (1991).

17. *See* Robert H. Frank, *If* Homo Economicus *Could Choose His Own Utility Function, Would He Want One With a Conscience?*, 77 AMER. ECON. REV. 593, 593–603 (1987) (providing a formal model).

18. *See* Saul Levmore, *Norms as Supplements,* 86 VA. L. REV. 1989, 1990–93 (2000).

19. *See* SHAVELL, *supra* note 13, at 603.

20. *See infra* text accompanying notes 80–84.

cient. For example, social and moral sanctions likely explain why there is widespread compliance with U.S. tax laws even though the odds that tax evasion will be detected and prosecuted are extremely small, and why airport nonsmoking rules and city pooper-scooper laws have strong behavioral effects even where they receive no legal enforcement at all.[21] Social and moral sanctions may even be more important than law to market efficiency. For example, the experience with simply adopting capitalist laws to create markets in former communist nations has had somewhat disappointing effects, which one might attribute at least in part to the lack of well-established social and moral business norms in those nations.[22] And the literature is replete with many other instances where behavior cannot be explained without the supplemental influence of social or moral norms.[23]

Such social and moral sanctions have a regulatory advantage when those imposing them are better informed about the particular actors and situation and can act in a more contextual way with lower procedural costs.[24] This is certainly true for moral sanctions. We each know what we did and can adjudicate that fact against ourselves with relative ease. Social sanctions also can be imposed at relatively low procedural cost, but those imposing them may be misinformed or inaccurate. Still, they often are imposed by those who are closer to the situation and thus more likely to know the true facts. Moreover, social sanctions are strongest when imposed by those whose views we care about, which usually means persons close and friendly enough to hear our side of the story and be relatively sympathetic to it.

Appropriate social and moral sanctions enable the legal system to reach a more optimal trade-off by narrowing laws or lowering penalties to reduce legal overdeterrence even when that creates greater legal underdeterrence because the legal system can rely on social and moral sanctions to reduce the latter problem. The legal system can adopt relatively low legal penalties because the legal violations that would otherwise result will be reduced by social and moral norms that encourage being law-abiding. The legal system can also be relatively underinclusive because the undesirable conduct that as a result lies outside legal prohibition still will be deterred by social and moral sanctions. These relatively low penalties and underinclusive laws will be socially desirable given the existence of social and moral sanctions, but they also will increase the importance of not permitting particular actors to insulate themselves from social and moral sanctions.

---

21. *See* Cooter, *Three Effects of Social Norms, supra* note 14, at 3–4, 10–11; Richard Craswell, *Do Trade Customs Exist?,* in The Jurisprudential Foundations of Corporate and Commercial Law 24–25 (Jody Kraus & Steven Walt, eds., 1997).

22. *See* Richard H. Pildes, *The Destruction of Social Capital Through Law,* 144 U. Pa. L. Rev. 2055, 2062–63 (1996) (summarizing literature).

23. *See* McAdams, *supra* note 13, at 340–41, 343–47.

24. *See* Cooter, *Three Effects of Social Norms, supra* note 14, at 21–22; Shavell, *supra* note 13, at 621–24.

Another advantage to social and moral norms is that often the right solution to a social problem is not the adoption of a law that mandates or forbids certain conduct for all persons. It is instead to have some, but not all, actors close to the scene provide some local service or benefits that they can supply more easily than government actors can. This results from the same problem of legal underinclusion because an unrealistically perfect legal system could fashion a legal rule or standard that would identify who the best local actors are in every situation. But the solution here is not to have a social or moral norm that also mandates or forbids conduct for an identified set of actors, just with better information. It is instead to have social or moral norms that induce charitable or volunteer impulses among enough of a broader set of local actors to produce the desired local service or benefit.

Whether moral and social sanctions improve behavior will depend on whether the underlying moral and social norms accurately identify undesirable behavior. If we live in a racist society, then moral and social norms might be designed to drive us to engage in undesirable racist behavior. But this problem is equally true of legal and economic sanctions. Whether they improve behavior depends on how accurately they identify undesirable conduct, and in a dysfunctional society, legal and economic sanctions may well encourage undesirable conduct. But generally speaking, moral and social sanctions, like legal and economic sanctions, will roughly reflect the views of the society that inculcated or created them. Of course, you or I may have different views than our society about which conduct is desirable, and probably do for at least some conduct. But if we agree with prevailing societal norms on enough conduct, then over the full range of conduct, social and moral sanctions would tend to move behavior in a direction we would find desirable. And even where this is not true, it is enough that social and moral sanctions would on balance advance the outcomes that our society views as desirable, which is the normative perspective relevant for determining what level of managerial discretion society should want to allow.

Another problem is that, like legal sanctions, social and moral sanctions might themselves be overinclusive and underinclusive.[25] But their regulatory advantages are likely to mean they would still improve the conduct that would result from a regime that used only legal and economic sanctions. Further, society can reduce the over- and underinclusion of norms with laws designed to expand the application of good norms and discourage the overapplication of bad ones.[26] Such legal regulation will, of course, itself be imperfect. But it seems reasonable to assume that on balance, the social and moral norms that attract other persons and are allowed to flourish by society do so because they improve behavior in the eyes of others and society.

---

25. *See* Shavell, *supra* note 13, at 607–08, 620–21; Eric A. Posner, *Law, Economics, and Inefficient Norms*, 144 U. Penn. L. Rev. 1697 (1996).

26. *See, e.g.*, McAdams, *supra* note 13, at 346–49, 391–432; Posner, *supra* note 25, at 1725–43; Shavell, *supra* note 13, at 618; Sunstein, *supra* note 13, at 907, 910, 947–65.

Conceivably, the judgment might go the other way. A society might determine that its own moral and social sanctions were overall counterproductive. If so, it then would make sense to impose, if feasible, a legal duty to profit-maximize on everyone, not just corporations, to override those sanctions. But as far as I know, no society has ever done so. Even if we assume away the enormous enforcement problems, it certainly would be a startling proposition, contrary to reams of moral philosophy, to conclude that our behavior is likely to be improved if each of us refused to consider the effects of our conduct on others unless it ultimately redounded to our own financial gain. The absence of any general duty for citizens to profit-maximize thus seems to reflect a revealed preference of society for allowing social and moral sanctions to operate.

Given this baseline system of social and moral regulation, the burden would seem to be on those advocating a duty of profit maximization for corporations to demonstrate that something special about corporations makes it desirable to prevent them from acting on the same social and moral impulses that help influence the conduct of noncorporate actors and businesses. That is the issue I address next, concluding that current law correctly finds no special rationale to impose such a special duty to profit-maximize on corporate managers. To the contrary, two important special features of corporations—shareholders' relative insulation from social and moral sanctions, and collective-action problems in acting on any social and moral impulses they have—make it particularly important to preserve managerial discretion to respond to social and moral considerations.

The Corporate Discretion to Refrain from Legal Profit-Maximizing Activity

Suppose there is no environmental regulation prohibiting clear-cutting, but it is nonetheless regarded as environmentally irresponsible. Can our corporate management decline to engage in clear-cutting even if it is in fact profit-maximizing? The legal answer is yes. Despite contrary assertions by advocates of a profit-maximization duty, the law has never barred corporations from sacrificing corporate profits to further public-interest goals that are not required by law.

As even proponents of a profit-maximizing duty concede, no corporate statute has ever stated that the sole purpose of corporations is maximizing profits for shareholders.[27] To the contrary, every state has enacted a corporate statute giving managers explicit authority to donate corporate funds for charitable purposes.[28] Because the argument for giving managers such donative discretion is actually weaker than the argument for giving them operational discretion, given that distributing profits to shareholders would allow them to make varying

---

27. *See* Clark, *supra* note 1, at 17, 678.

28. *See* Choper, Coffee & Gilson, Cases and Materials on Corporations 39 (6th ed. 2004) [hereinafter "Choper, Coffee & Gilson 6th Edition"].

donative decisions but would not enable them to make varying operational decisions, this suggests the state legislatures also would favor the latter. But we don't have to guess about that because 30 states have adopted corporate constituency statutes that explicitly authorize managers to consider nonshareholder interests, specifically including the interests not only of employees, but also of customers, suppliers, creditors, and the community or society at large.[29] Although these constituency statutes were prompted by the 1980s takeover wave, most are not limited to takeovers, but apply to any management decision.[30]

Even without any statute, such discretion has been recognized by the corporate common law that governs absent statutory displacement. The influential Principles of Corporate Governance by the American Law Institute (ALI) state that without any statute, the basic background rule regarding for-profit corporations is as follows:

> Even if corporate profit and shareholder gain are not thereby enhanced, the corporation, in the conduct of its business ... (2) May take into account ethical considerations that are reasonably regarded as appropriate to the responsible conduct of business; and (3) May devote a reasonable amount of resources to public welfare, humanitarian, educational, and philanthropic purposes.[31]

The ethics provision plainly gives managers operational discretion to sacrifice profits to avoid conduct that might "unethically" harm employees, buyers, suppliers, or communities.[32] Depending on how elastic one's conception of ethics is, that could cover the bulk of socially responsible conduct. Any profit-sacrificing public-spirited activity not covered by the ethics provision would seem covered by the next one, which the ALI comments make clear authorizes not just donations, but also operational decisions such as declining to make profitable sales that would adversely affect national foreign policy, keeping an unprofitable plant open to allow employees to transition to new work, providing a pension for former employees, or other decisions that take into account the social costs of corporate activities.[33] Likewise, the ALI rule on hostile takeovers explicitly states that "in addition to" considering shareholder interests and economic prospects, the board can consider "interests or groups (other than shareholders) with respect to which the corporation has a legitimate concern if to do so would

---

29. American Law Institute, Principles of Corporate Governance: Analysis and Recommendations §2.01, Reporter's Note 8, §6.02, Comment a (1994) [hereinafter "ALI Principles"].

30. *See* ALI Principles, *supra* note 29, §2.01, Reporter's Note 8; ABA, *supra* note 1, at 2266.

31. ALI Principles, *supra* note 29, §2.01(b)(2)–(3) & Comment d.

32. ALI Principles, *supra* note 29, §2.01 Comment h.

33. *See* ALI Principles, *supra* note 29, §2.01 Comment i & Illustrations 13, 20, 21. Indeed, the stronger the nexus to corporate operations, the more likely the decision to sacrifice profits would be sustained under 2.01(b)(3). *Id.*

not significantly disfavor the long-term interests of shareholders."[34] This necessarily authorizes blocking takeovers that would sacrifice some amount of long-term shareholder profits.

True, even though they generally endeavor to restate existing law, the ALI provisions are not themselves legally binding. But the ALI provisions do cite case law supporting these provisions that go beyond the proposition that such public-spirited corporate conduct is permissible when it happens to maximize corporate profits in the long run.[35] Even the supposedly conservative Delaware, which does not have a corporate constituency statute, does by case law authorize managers to reject a takeover bid based on "the impact on 'constituencies' other than shareholders (i.e., creditors, customers, employees, and perhaps even the community generally), [and] ... [w]hile not a controlling factor ... the basic stockholder interests at stake."[36] Delaware case law also holds that managers may rebuff tender offers based on "any special factors bearing on stockholder *and* public interests."[37] Federal courts have similarly construed the state corporate laws of numerous other states.[38] And even pretakeover Delaware case law explicitly held that managers could make donations that sacrificed a "reasonable" amount of shareholder profits to further public-interest objectives.[39] True, when corporate control is being sold, then that does trigger a duty to profit-maximize for special reasons that I discuss later. But we will see that the cases so holding emphasize that this profit-maximization duty applies only to such sales of corporate control, and thus make clear that it does not apply otherwise.

Proponents of a profit-maximization duty often try to narrow these contrary provisions and case law by reading them to authorize making donations, being ethical, and considering nonshareholder interests only to the extent that doing

34. ALI Principles, *supra* note 29, §6.02(b)(2). *See also id.* Comment c(2) ("Such groups and interests would include, for example, environmental and other community concerns, and may include groups such as employees, suppliers, and customers.")

35. ALI Principles, *supra* note 29, §2.01 Reporter's Note 2. *See also* Herald Co. v. Seawell, 472 F.2d 1081, 1091 (10th Cir. 1972).

36. Unocal Corp. v. Mesa Petroleum Co., 493 A.2d 946, 955 (Del. 1985). *See also* Ivanhoe Partners v. Newmont Mining, 535 A.2d 1334, 1341–42 (Del. 1987) ("the board may under appropriate circumstances consider ... questions of illegality, the impact on constituencies other than shareholders, ... and the basic stockholder interests at stake"); Paramount Communications v. Time, 571 A.2d 1140, 1153 (Del. 1990) ("directors may consider, when evaluating the threat posed by a takeover bid, ... 'the impact on "constituencies" other than shareholders.'").

37. Mills Acquisition v. Macmillan, Inc., 559 A.2d 1261, 1285 n.35 (Del. 1989) (emphasis added). The Court also there stated that managers may base their rejection on the "effect on the various constituencies, particularly the stockholders," which implicitly indicates that the analysis is not limited to the effect on shareholders. *Id.*

38. *See* I Dennis J. Block, et al., The Business Judgment Rule 809–10 (5th ed. 1998) (collecting cases); GAF Corp. v. Union Carbide Corp., 624 F. Supp. 1016, 1019–20 (S.D.N.Y. 1985).

39. *See* Theodora Holding Corp. v. Henderson, 257 A.2d 398, 405 (Del. Ch. 1969). *See also* Kahn v. Sullivan, 594 A.2d 48, 61 (Del. S.Ct. 1991), (same).

so maximizes profits in the long run.[40] But this narrow reading is strained. Nothing in the language of the ALI or statutory provisions limits them to cases where there is a convenient coincidence between maximizing profits and the public interest. To eliminate any doubt, the ALI comments explicitly state that these provisions apply "even if the conduct either yields no economic return or entails a net economic loss."[41] The ALI comments also clearly stress that although the conduct covered by these provisions is often profit-maximizing in the long run, these provisions authorize such conduct even when that isn't true.[42]

Likewise, the corporate constituency statutes generally have separate provisions, one stating that managers may consider the long- and short-term interests of shareholders, and another that managers can consider the effects of corporate conduct on other constituencies.[43] The latter provision would be superfluous if it merely allowed managers to consider those effects when they had an impact on the long- or short-term interests of shareholders. Thus the standard canon of statutory construction that, where possible, a statute should be interpreted to render all provisions meaningful indicates that the latter provision must have been intended to allow consideration of the impact on those other constituencies even when it did not maximize long- or short-term shareholder interests. Some states' statutes even explicitly reject the proposition that management must regard the interests of any particular group such as shareholders "as a dominant or controlling interest or factor"; one has an official comment saying that the statute "makes clear that a director is not required to view presently quantifiable profit maximization as the sole or necessarily controlling determinant of the corporation's 'best interests.'"[44] Another state's statute expressly authorizes managers to decide that "a community interest factor ... outweigh[s] the financial or other benefits" to shareholders.[45]

It is also hard to believe legislatures would have bothered to enact corporate constituency statutes simply to affirm the ability of managers to consider factors that might increase shareholder profits. The "no change" interpretation of these statutes by profit-maximization proponents also seems inconsistent with the

40. Clark, *supra* note 1, at 682–683; ABA, *supra* note 1, at 2269; *see also* Block et al., *supra* note 38, at 810–811, 813–14 (reporting some efforts to do so).

41. ALI Principles, *supra* note 29, §2.01, Comment f. *See also id.* §4.01, Comment d to §4.01(a), first paragraph ("There are, of course, instances when § 2.01 would permit the corporation to voluntarily forgo economic benefit—or accept economic detriment—in furtherance of stipulated public policies ... ethical considerations ... or public welfare, humanitarian, educational, or philanthropic purposes").

42. ALI Principles, *supra* note 29, §2.01, Comments h & i. Indeed, the Comments authorize some degree of such conduct when it could not maximize long-run profits because the company is liquidating and thus has no long run. *See id.* Illustration 13.

43. *See, e.g.,* N.Y. Bus. Corp. Law §717(b); Conn. St. §33-756(d).

44. *See, e.g.,* Pa. Bus. Corp. Law §1715(b); Pa. Corp. Law §515(b); Ind. Code Ann. §23-1-35-1(f) & Official Comment to (d).

45. *See* Iowa Code Ann. §491.101B.

fact that these proponents vociferously oppose these statutes as well[46]—if the statutes just identify factors relevant to figuring out what maximizes profits, what's the beef?

The state corporate statutes authorizing charitable donations indicate a similar conclusion. Twenty-four states, including Delaware, authorize "donations for the public welfare or for charitable, scientific, or educational purposes,"[47] which is similar enough to the last ALI provision to suggest a similar power to sacrifice profits. Further, 19 other corporate statutes, as well as the Revised Model Business Corporation Act, make this even clearer by having *separate* provisions, one authorizing donations "furthering the business and affairs of the corporation," and another one authorizing, as in the first 24 states, "donations for the public welfare or for charitable, scientific, or educational purposes."[48] The first provision would render the latter provision superfluous if the latter authorized only to donations that furthered the business of the corporation. Thus, again, the canon that a statute should be interpreted to render all provisions meaningful governs, and here implies that the latter sort of provision must authorize donations for charitable and public-welfare purposes that do *not* further the business and affairs of the corporation. The remaining seven states, which include our biggest ones, California and New York, are the most explicit of all, authorizing charitable donations "irrespective of corporate benefit."[49] Further, although corporate managers generally claim their donations increase long-run profits, as an empirical matter this frequently seems dubious;[50] thus, in fact, profit-sacrificing donations are being allowed.

Federal law also seems to recognize a discretion to sacrifice corporate profits to further public-interest objectives because Rule 14a-8 allows shareholder proposals on social responsibility issues significantly related to the corporation's businesses, even when not motivated by profit-maximizing concerns. As the SEC made clear in adopting this amendment, and as subsequent cases have held, this includes proposals whose significance in relation to corporate business is ethical rather than financial.[51]

None of this means that managers have a legally enforceable *duty* to engage in profit-sacrificing conduct when not required by other law to do so. The above legal authorities all use language of discretion. The Connecticut corporate constituency statute might seem the exception, for it does contain mandatory language that managers "shall consider" various nonshareholder interests.[52] But

46. *See* ABA, *supra* note 29, at 2253, 2268.

47. *See* Choper, Coffee & Gilson 6th Edition, *supra* note 28, at 39.

48. *See id.*; RMBCA §3.02(13)&(15).

49. *See* Choper, Coffee & Gilson 6th Edition, *supra* note 28, at 39.

50. *See* Victory Brudney & Allen Ferrell, *Corporate Charitable Giving*, 69 U. Chic. L. Rev. 1191, 1192 n.4, 1195 (2002).

51. SEC Exch. Act Rel. No. 34-12,999 (1976); Lovenheim v. Iroquois Brands Ltd., 618 F. Supp. 554 (D.D.C. 1985).

52. Conn. Stat. §33-756(d).

because this statute at most sets forth a duty to the corporation, it can be enforced against managers only via a derivative action by shareholders. Thus nonshareholder interests have no way of forcing managers to even consider their interests if managers prefer not to, though an interesting case could arise if they bought some shares in a Connecticut corporation in order to do so. In any event, even if managers of Connecticut corporations did have a truly enforceable duty to consider nonshareholder interests, nothing in the law requires them to give those interests any particular weight, so their discretion remains undisturbed.

Proponents of a profit-maximization duty generally rely on the duty of care, which in most states provides that a manager should discharge his duties "in a manner he reasonably believes to be in the best interests of the corporation."[53] But duty-of-care laws never define the "best interests of the corporation" as meaning solely the interests of shareholders, nor do they ever define the interests of shareholders to mean solely their *financial* interests. Both are glosses added by proponents. Indeed, as noted above, corporate constituency statutes in most states explicitly reject that definition by providing that in evaluating the "best interests of the corporation," a director may consider the effects of corporate action on shareholders, employees, suppliers, customers, or the larger community. The comments to the ALI Principles explicitly state that acts that "voluntarily forgo economic benefit—or accept economic detriment—in furtherance of stipulated public policies ... ethical considerations ... or ... public welfare, humanitarian, educational, or philanthropic purposes ..., even though they may be inconsistent with profit enhancement, should be considered in the best interests of the corporation and wholly consistent with [duty-of-care] obligations."[54] And as noted above, courts have explicitly sustained profit-sacrificing corporate decisions despite the duty of care.

In any event, even if the duty of care did nominally require profit maximization, the business judgment rule makes plain that it cannot be enforced in a way that would bar managers from exercising discretion to sacrifice corporate profits in the public interest. Under the business judgment rule, the courts will not second-guess managers' business judgment about what conduct is in the best interests of the corporation unless those managers have a conflict of interest. Statutes define conflicts of interest to include only "the financial interests of the director and his immediate family and associates,"[55] thus making clear that this exception does not apply if the alleged conflict is between the corporation's financial interests and some public-interest cause, even if the manager derives a special

53. *See* Clark, *supra* note 1, at 679 & n.2; MBCA §8.30(a); ALI Principles §4.01(a) & Reporter's Note 1.

54. ALI Principles, *supra* note 29, §4.01, Comment d to §4.01(a), first paragraph.

55. ABA Committee on Corporate Laws, *Changes in the Model Business Corporation Act—Amendments Pertaining to Directors' Conflicting Interest Transactions*, 43 Bus. Law. 691, 694 (1988) [hereinafter, ABA, *Conflicting Interest Transactions*]; *see also* Conn. Stat. §33-781; 8 Del. Code §144(a); N.Y. Bus. Corp. §713(a); Cal.Corp.Code § 310(a); Revised Model Bus. Corp. Act §8.60.

psychic pleasure from furthering it. Moreover, in applying the business judgment rule, courts refrain from reviewing not only whether the conduct actually increased profits, but also whether it was seriously likely to do so, or even whether managers were actually motivated by profit maximization when they exercised their judgment. The result is that, under the business judgment rule, courts are extraordinarily willing to sustain decisions that apparently sacrifice profits, at least in the short run, on the ground that they may conceivably maximize profits, at least in the long run. Because just about any decision to sacrifice profits has a conceivable link to long-term profits, this suffices to give managers substantial de facto discretion to sacrifice profits in the public interest.

Illustrative is *Shlensky v. Wrigley*,[56] which considered a claim that the directors of the corporation owning the Chicago Cubs (including 80 percent shareholder Mr. Wrigley) had violated their fiduciary duties by refusing to install lights in Wrigley Field. The complaint alleged that Mr. Wrigley "has admitted that he is not interested in whether the Cubs would benefit financially" from installing lights, but rather was motivated by "his personal opinions 'that baseball is a "daytime" sport and that the installation of lights and night baseball games will have a deteriorating effect upon the surrounding neighborhood.'"[57] The complaint further alleged a plethora of facts supporting a conclusion that installing lights would in fact have increased corporate profits: (1) every other baseball team had installed lights for the purpose of increasing attendance and revenue; (2) Cubs road attendance, where night baseball was played, was better than Cubs home attendance; (3) Cubs weekday attendance was worse than that of the Chicago White Sox, who played at night in the same city, even though their weekend attendance, when both teams played day ball, was the same; and (4) the cost of installing lights, which could be financed, would be more than offset by the extra revenue that would result from increasing attendance by playing night baseball.[58]

The court affirmed dismissal of the complaint, stating that it was "not satisfied that the motives assigned to [Mr. Wrigley] are contrary to the best interests of the corporation and the stockholders," because in the long run a decline in the quality of the neighborhood might reduce attendance or property value.[59] But the court did not allow inquiry into whether such long-run profitability was Mr. Wrigley's actual motivation. Rather, it held irrelevant any motives other than fraud, illegality, or conflict of interest, thus rendering moot the allegations that Mr. Wrigley was motivated not by corporate profits, but by public-interest concerns.[60] Nor did the court hinge its holding on any conclusions about whether

56. 95 Ill. App. 2d 173, 237 N.E.2d 776 (1968).
57. 95 Ill. App. 2d at 176.
58. *Id.* at 175–76.
59. *Id.* at 180–81.
60. *Id.* at 181.

continuing day baseball would actually maximize profits, or was at all likely to do so, saying that such matters were "beyond our jurisdiction and ability."[61]

Thus even if profit maximization were the nominal standard, business judgment review still would sustain any public-spirited activity without any inquiry into actual profitability or the manager's actual purposes as long as it had some conceivable relationship to long-run profitability, however tenuous.[62] And such a relationship can almost always be conceived. Indeed, it is hard to see what socially responsible conduct could not plausibly be justified under the commonly accepted rationalizations that it helps forestall possible adverse reactions from consumers, employees, the neighborhood, other businesses, or government regulators—especially given that managers know they need not "particulariz[e]" any profits they claim this reaps.[63] Because such business judgment review suffices to sustain the lion's share of decisions to sacrifice corporate profits in the public interest, courts rarely need to state explicitly that managers have such discretion.

The contrary case on which profit-maximization proponents tend to focus is the 1919 case *Dodge v. Ford Motor,*[64] but that old precedent does not really support the proponents' claim. In that case, the Dodge brothers, 10 percent shareholders in Ford Motor Company, sued because Henry Ford and his fellow directors had stopped paying special dividends to shareholders in order to fund an expansion of operations that would allow the firm to increase employment and cut prices.[65] The decision did include some strong pro-shareholder profits language and required Ford Motor to distribute more of its profits in dividends. But the opinion never stated that directors' exclusive duty is to maximize shareholder profits. Rather, it held that profits should be the primary but not exclusive goal of managers, and sustained the expansion despite its factual conclusion that management based that operational decision largely on humanitarian motives.

The *Dodge* court stated: "We do not draw in question, nor do counsel for the plaintiffs do so, the validity of the general proposition stated by counsel ... [that] '[a]lthough a manufacturing corporation cannot engage in humanitarian works as its principal business, the fact that it is organized for profit does not prevent the existence of implied powers to carry on *with humanitarian motives* such charitable works as are incidental to the main business of the corporation.'"[66] The court also said that "an incidental humanitarian expenditure of corporate funds for the benefit of the employes [*sic*]" was permissible. Thus, if "incidental" to business operations, an expenditure could be for the benefit of charities or employees, rather than for the ultimate benefit of shareholders.[67]

The court accordingly made clear that corporate conduct did *not* have to have the ultimate aim of increasing long-run shareholder profits. Instead, what the

61. *Id.* at 181.

62. Clark, *supra* note 1, at 682–83.

63. ALI Principles, *supra* note 29, §2.01, Comment f & Illustrations 1–5.

64. 204 Mich. 459, 170 N.W. 668 (1919); *see. e.g.*, Clark, *supra* note 1, at 679.

65. 204 Mich. at 464, 503–07.

66. *Id.* at 506 (emphasis added).

67. *Id.* at 506.

court emphasized was that the discretion to do otherwise was *bounded* by a requirement that other purposes remain incidental to a primary purpose of profiting shareholders. It stated that corporations are organized "primarily" for the shareholder profits, and thus cannot "conduct the affairs of a corporation for the merely incidental benefit of shareholders and for the primary purpose of benefitting others."[68] This language limits the degree of profit-sacrificing discretion rather than imposing a duty to exclusively profit-maximize.

Further, in terms of actual discretion, what matters is less such general language as what the courts actually sustain. And the *Dodge* court in fact sustained the directors' operational decisions, refusing to enjoin the expansion and expressly noting directors had discretion over pricing.[69] It did so even though the court held that this business plan would clearly reduce short-run profits, and that the court had "no doubt that certain sentiments, philanthropic and altruistic, ... had large influence in determining the policy to be pursued by the Ford Motor Company."[70] The court reasoned that it was "not satisfied that the alleged motives of the directors, in so far as they are reflected in the conduct of the business, menace the interests of the shareholders" because "the ultimate results of the larger business cannot be certainly estimated" and "judges are not business experts."[71] Thus the court was not willing to strike down the conduct based on the managers' actual subjective motive as opposed to what could be inferred from their conduct. Nor was the court willing to assess the actual profitability of the conduct. Instead, the court sustained the conduct on the grounds that it could conceivably be profitable in the long run. This suffices to confer considerable de facto discretion even if one did wrongly interpret the opinion to impose a nominal duty of pure profit maximization.

The *Dodge* court did strike down the refusal to declare any special dividends, but that was because the court found that the corporation was withholding more than it needed to fund the expansion.[72] Thus the refusal to declare special dividends was stricken not because it had a public-interest motive, but because it went beyond any business *or* public-interest motive. What then could have been the purpose of withholding unneeded funds? Most likely it was that suspending dividends would depress stock prices, and thus force the Dodge brothers to sell their stock to majority shareholder Henry Ford at favorable prices, which eventually happened; this would violate Henry Ford's fiduciary duty not to use his corporate control to benefit himself financially at the expense of other shareholders.[73] That is, the otherwise aberrational court decision to interfere with the exercise of management discretion about dividend levels seems best explained on the view that the case really involved a conflict of interest raising duty-of-loy-

68. *Id.* at 507.

69. *Id.* at 507–08.

70. *Id.* at 504–06.

71. *Id.* at 508.

72. *Id.* at 508–09.

73. *See* Clark, *supra* note 1, at 604; D. Gordon Smith, *The Shareholder Primacy Norm,* 23 J. Corp. L. 277, 315, 318–20 (1998).

alty concerns. In any event, the decision on dividends involved no actual sacrifice of profits, but a choice about whether to hold or distribute those profits. Thus the court order did not actually interfere with any management decision to sacrifice profits in the public interest.

So even *Dodge*, the high-water mark for the supposed duty to profit-maximize, indicates that no such enforceable duty exists. Nor do there appear to be any other cases that have actually restrained a management decision to sacrifice corporate profits in the public interest. Rather, the cases uniformly sustain profit-sacrificing conduct by either (1) using lax business judgment review standards to accept strained claims of conceivable long-run profitability or (2) concluding that reasonable amounts of profit sacrificing are legal.[74]

Why an Operational Discretion to Sacrifice Corporate Profits in the Public Interest Is Desirable and Even Efficient

Even if the law gives management the discretion to sacrifice corporate profits in the public interest, is that desirable? The answer turns out to be yes, and more surprisingly, it is yes even if we assume that economic efficiency is our ultimate metric of desirability. This is true even if we wrongly equate efficiency with shareholder profit maximization. And it is even more clearly true once we recognize that shareholders have interests other than economic ones, and that the corporate structure has implications for the ability of social or moral sanctions to police corporate conduct that might inefficiently harm those outside the corporation.

Why Even a Legal Regime that Maximizes Shareholder Profits Necessarily Confers Managerial Discretion to Sacrifice Profits in the Public Interest

Even if one narrowly (and mistakenly) defined efficiency to equal shareholder profit maximization, managerial discretion to sacrifice profits is still necessary because the economic efficiencies that come from delegating management to someone other than shareholders or judges cannot be achieved without creating such discretion. As economists have shown, the optimal level of agency costs requires some trade-off between monitoring costs and the costs of permitting agent discretion even if one assumes shareholder profitability is the only goal.[75] In the economic lingo, giving such discretion to managers lowers total agency costs because any residual loss of shareholder profits is offset by the savings in monitoring costs, which we might equally call the benefits of delegation.

---

74. *Accord* ALI Principles, *supra* note 29, §2.01 Reporter's Note 2; A.P. Smith Mfg. Co. v. Barlow, 13 N.J. 145, 150–54, 161 (1953).

75. *See* Jensen & Meckling, *Theory of the Firm: Managerial Behavior, Agency Costs and Ownership Structure*, 3 J. Fin. Econ. 305, 308 (1976).

As a result, the economically efficient level of agency costs will always leave some agency slack: that is, some agent discretion to act in ways other than in the financial interests of the shareholders. And the agents who can exercise such agency slack to sacrifice corporate profits by benefiting themselves (say by renting corporate luxury boxes in stadiums) can also do so by benefiting the public interest (say by not clear-cutting). In either case, shareholders focused on the bottom line will care about only the total amount of agency slack and profit-sacrificing behavior, and not about precisely how those profits were sacrificed. And in either case, a strained claim that the activity somehow increases corporate profits (by building goodwill with clients or the community) will mean that it will survive legal scrutiny under the business judgment rule, which sets what both the law and proponents of a duty to profit-maximize regard as the optimal degree of legal monitoring. As we have already seen, this business judgment rule level of monitoring effectively eliminates any enforceable duty to profit-maximize and leaves managers with de facto discretion to sacrifice a reasonable degree of corporate profits to further public-interest objectives.

Understanding this point neatly deflates the argument by proponents of a duty to profit-maximize that the goal of profit maximization is objective and easier to monitor than a goal of advancing the public interest, which, because it goes beyond legal compliance, is either vague or controversial.[76] To begin with, the ability of judges to monitor public-interest goals is irrelevant because the claim is not that corporate managers should have some ill-defined legal duty to pursue the public interest; the claim is that they have discretion to do so, in part because the business judgment rule inevitably gives it to them. In contrast, a real enforceable duty to profit-maximize would require judicial monitoring, and thus runs against the problem that the very reason for the business judgment rule is precisely that profit maximization is too *hard* for judges to monitor.

It seems dubious that even the most energetic judicial efforts to force corporate managers to maximize profits at the expense of nonprofit goals would be at all effective. After all, if we thought judges were better than managers at figuring out what maximizes corporate profits, then why have corporate managers at all rather than allow judges to make all corporate decisions? Presumably shareholders instead delegate managerial authority to professional managers because they are better at managing businesses than judges are. Indeed, judges themselves have repeatedly expressed their lack of expertise in gauging the profitability of managerial decisions.[77] Any more vigorous judicial enforcement would thus likely increase the error rate, with mistaken judicial decisions (or the risk of them) deterring business decisions that actually would have increased shareholder profits.

One might imagine judges instead focusing more on managers' actual motives, rather than abjuring such an inquiry as the current doctrine does. But

---

76. *See, e.g.*, CLARK, *supra* note 1, at 20, 679, 692, 702; ABA, *supra* note 1, at 2269–70.

77. See, for example, the above quotes from *Wrigley* and *Dodge*.

such motivational inquiries are problematic for all the familiar reasons,[78] including the imponderable difficulties of sorting out mixed motives. More important, because subjective motives are unobservable (absent the rare and unlikely to recur case of an explicit admission by management), courts will have to ascertain motivation based on which purposes seemed objectively likely given the observable circumstances. And that inevitably will turn on the court's business judgment about which method of operation actually would maximize profits, which again gets us into the problem that courts are worse at making such decisions than corporate managers are.

Moreover, even where motives are clear, proving damages or an actual injury justifying injunctive relief would necessitate a causation inquiry that would require the court to make a business judgment as to whether a different method of operations actually would have led to more profit. Even if putting lights in Wrigley Field would have created profits that exceeded all costs, including costs of operation and maintenance, how could a court ever decide whether investing those funds in another relief pitcher would have produced even more profit? Avoiding such motivation and causation inquiries would seem advisable for all the reasons underlying judicial deference to business judgment.

These difficulties with judicial enforcement of a duty to profit-maximize are only worsened in situations where alleged profit sacrifices advance public-interest objectives. As the discussion of the *Wrigley* and *Dodge* cases indicates, courts have a hard time figuring out whether corporate conduct really sacrificed profits in the short run at all. And even if courts could figure that out, courts have no real way of assessing the conventional claim that those short-run profit losses are offset by the long-run profits that result because the conduct produces goodwill that in turn increases sales, employee efforts, local property values, or favorable treatment by other businesses, the community, or the government. Such claims are often dubious, but given their nature and counterfactual quality, they are less likely to rest on hard data than on intuitive judgments that managers are better placed to make than judges. Further, they require judgments about the correct discount rate to apply to future profits, about which courts lack not only expertise, but any governing legal principle.[79]

Worse, this commonly understood problem actually understates the difficulty. Even greater difficulties are raised by possible disjunctions between ex post and ex ante profit maximization. Proponents of a profit-maximization duty normally seem to assume that if decision X will maximize the combination of short- and long-term profits at the time the decision is made, then the manager must make decision X. But suppose, as many scholars have argued, that a manager with discretion not to make decision X sometimes can enter into an implicit contract that she won't do X in exchange for others (say workers or the community) con-

78. *See, e.g.*, Clark, *supra* note 1, at 137–38.

79. *See* Paramount Communications v. Time, 571 A.2d 1140, 1154 (Del. 1990) (holding that managers must determine the right time frame); Henry Hu, *Time, Risk, and Fiduciary Principles in Corporate Investment*, 38 UCLA L. Rev. 277, 301–02 (1990).

ferring some benefit on the corporation that cannot be taken back (say harder work or favorable zoning review), and that such implicit contracts often are more profitable and efficient than legally binding commitments would be.[80] For example, suppose it is profit-maximizing for a corporation to enter into an implicit contract with its employees that they will work to develop their skills in a way that makes them more valuable to the corporation than to other firms, in exchange for the corporation refraining from cutting their salaries to levels that do not reflect their firm-specific investments of human capital. In that sort of case, the later decision to refrain from doing X (cutting salaries) would look profit-sacrificing from an ex post perspective that considers only postdecision profits, but would be ex ante profit-maximizing when one considers that the ability to make that later decision was necessary to create a profitable implicit contract. Allowing managers to exercise discretion to sacrifice ex post profits in such a case thus enables them to enter into implicit contracts that are ex ante profit-maximizing. Lacking legal enforcement, such implicit contracts must owe their enforcement to social or moral sanctions against reneging on such loose understandings, which can be effective only if not overridden by a legal duty.

As Professors Blair and Stout have noted, this point is not limited to implicit contracts that require some special understanding between the corporation and others, but can justify the general existence of manager profit-sacrificing discretion on the ground that it is likely to reward and thus encourage firm-specific investments by other stakeholders that are ex ante profit-maximizing.[81] With the analysis in this article, we can further develop this point to say that the mere existence of profit-sacrificing discretion can be ex ante profit-maximizing because of a very general expectation that such discretion will make managers responsive to social and moral sanctions.[82] Suppose that others will comply with social or moral norms that are beneficial to the corporation only on the expecta-

80. Early work in this vein argued that takeovers might cause implicit contract breaches that were profitable ex post but decreased corporate profits ex ante by discouraging other stakeholders from making firm-specific investments. *See* John C. Coffee, Jr., *The Uncertain Case for Takeover Reform: An Essay on Stockholders, Stakeholders and Bust-ups,* Wis. L. Rev. 435, 446–48 (1988); John C. Coffee, *Shareholders versus Managers,* 85 Mich. L. Rev. 1, 9, 23–24, 73–86 (1986) [hereinafter Coffee, *Shareholders v. Managers*]; John C. Coffee, *Regulating the Market for Corporate Control,* 84 Colum. L. Rev 1145, 1234–44 (1984); Charles R. Knoeber, *Golden Parachutes, Shark Repellents, and Hostile Tender Offers,* 76 Am. Econ. Rev. 155, 161 (1986); Andrei Shleifer & Lawrence Summers, *Hostile Takeovers as Breaches of Trust,* in Corporate Takeovers 33 (A. Auerbach, ed. 1988). Professors Blair and Stout have since generalized this point beyond takeovers to argue that it justifies a general manager discretion to favor other stakeholders to reward and encourage their firm-specific investments. *See* Margaret M. Blair & Lynn A. Stout, *A Team Production Theory of Corporate Law,* 85 Va. L. Rev. 247, 304–05 (1999).

81. *See* Blair & Stout, *supra* note 80, at 275, 285.

82. See text accompanying notes 17–20. The point is similar to Robert Frank's model showing that merely having a trustworthy character will induce others to enter into efficient transactions with us, thus making it efficient to commit to having such a character, even when it is disadvantageous to us. *See* Frank, *supra* note 17, at 593–603. Here, the commitment is to have managers who are free to act based on their character even when that becomes unprofitable.

tion that the corporation will comply with social and moral norms that are beneficial to others. For example, suppose a town will comply with social and moral norms not to exact all it can from a corporation on a zoning issue but only because it expects the corporation to comply with social and moral norms to avoid some profit-maximizing environmental harms. The town has no special understanding with the corporation; its expectations simply affect whether it calculates that the corporation will confer a net benefit on the town. In this sort of case, the mere fact that managers have the discretion to engage in ex post profit-sacrificing compliance with social and moral norms (here by avoiding certain environmental harms) would be ex ante profit-maximizing. Shareholder profits would then be increased by a regime that gave corporate managers discretion to engage in such ex post profit sacrifices, for the prospect of such managerial behavior would encourage others to treat the corporation in beneficial ways that increased profits before the profit-sacrificing decision had to be made. This requires no special understanding of the sort that one might call an implicit contract; just a very general sort of social understanding that actors are likely to comply with social and moral norms, which leads to a social reciprocity that is profit-maximizing for each actor.[83] A duty to profit-maximize ex post ironically would decrease shareholder profits by constraining this discretion and thus disable the corporation from engaging in such profit-maximizing social reciprocity.

Such a claim of ex ante profit maximization was recognized implicitly by the supposedly conservative *Dodge* opinion, which stated that it did not doubt the soundness of a prior U.S. Supreme Court case that had sustained a decision by corporate managers to give away the corporation's water to a city for municipal uses, where the city had previously given the corporation the rights to lay its pipes and carry water to residents, and most stockholders resided in that city.[84] One might try to shoehorn this case into a story about long-term future profits by speculating that if the corporation had not given the city water, the city might have tried to take away the corporation's pipe rights or otherwise exacted regulatory vengeance. But any such subsequent city effort would have faced considerable problems under the taking clause because the corporation would have had vested property rights. In any event, the Court did not rely on any such theory, nor was any evidence cited of an implicit contract or special understanding at the time the corporation got the pipe rights. Instead, the Supreme Court sustained the corporate conduct as a proper means of reciprocating for the past (and

83. Because this theory requires profit-maximizing social reciprocity, it does not suffice to explain all corporate compliance with social and moral norms, or even compliance with all norms that are collectively profit-maximizing but individually unprofitable. For such norms, deviations can be profitable, even if others expect deviations in advance, if collective-action problems deprive any other individual actor of incentives to reciprocate in ways that decrease the individual corporation's profits. Enforcement through nonfinancial social or moral sanctions thus will be necessary, as discussed above in the section on social regulation of conduct.

84. 204 Mich. at 506 (citing Hawes v. Oakland, 104 U. S. 450, 461–62 (1881)).

literally sunk) benefits the city had conferred by allowing the underground pipes. One could restate this conclusion in the modern economic lingo by concluding that the city decision to award these profitable piping rights to the corporation might never have been made without the prospect that corporate managers would have the discretion to comply with social or moral norms of gratitude (here by engaging in future profit sacrifices to reward the city for that favorable treatment when the city needed it), so that sustaining the discretion to give away water was ex ante profit-maximizing even though it diminished the stream of ex post profits that followed the water giveaway.

Thus any duty to profit-maximize theoretically and legally would have to admit the defense that, even if the managerial decision sacrificed profits, the prospect that management would have the discretion to make such profit-sacrificing decisions encouraged others to treat the corporation well in ways that increased prior profits and thus made that discretion ex ante profit-maximizing. The problem, of course, is that although such a defense is conceptually valid, there appears to be no way for courts to reliably ascertain when it is true in specific cases. Even when there is a special understanding that rises to the level of an implicit contract, a defining feature of such contracts is that they aren't written down, making it difficult to verify their existence or terms. Even more difficult to verify would be the looser social understanding that others have greater incentives to treat the corporation well if they expect that social and moral sanctions will induce better behavior by the corporation in the future. Thus even if courts could overcome the insuperable problems involved with figuring out whether a managerial decision actually sacrificed *post*-decision profits, courts would have no way of really enforcing a legal duty to profit-maximize in the face of theoretical claims that managerial discretion to make decisions that sacrificed *post*-decision profits actually maximized shareholder profits ex ante.

It may be that profit maximization is an easier goal for *shareholders* to monitor than public-interest objectives. But declining to make profit maximization an enforceable legal claim does nothing to prevent shareholders from choosing to adopt profit maximization as the goal they choose to monitor in exercising their voting or investment rights. It simply means that dissenting shareholders cannot expect courts to enforce profit maximization over other goals unrelated to the personal financial interests of managers. Although shareholder monitoring is inevitably imperfect, shareholders are likely to be better than judges at making the sort of nuanced judgments about profitability required by the kinds of issues discussed above. Or, more likely, shareholders won't get into any details, but will just monitor the overall profitability of the firm to make sure managers do not come below profit expectations. This shareholders can do in an ongoing fashion more effectively than the litigation process, which is notoriously slow and after-the-fact.

In short, even if shareholder profit maximization were our only goal, fulfilling it inevitably would create considerable management discretion to sacrifice profits in the public interest. True, this theory explains only the latent discretion to sacrifice profits in the public interest that inevitably results from the business judgment rule itself. It does not provide the sort of affirmative justification that would explain why the law goes beyond that to allow even patent exercises of discretion that do not pretend to maximize profits either ex post or ex ante. For that, we need a more affirmative justification for the desirability of sacrificing corporate profits in the public interest, to which I turn in the next section.

But the inevitable existence of this latent discretion, even if one favors profit maximization, remains enormously important because, in the lion's share of cases, it produces the same result as a limited patent profit-sacrificing discretion. This means that both the existence and degree of profit-sacrificing discretion is largely inevitable. It also means that the fact that the law has taken the next step of embracing, when necessary, a limited patent discretion to sacrifice profits in the public interest produces little, if any, reduction in profits. The smallness of this marginal reduction in profits makes it easier to justify with any affirmative gains from the managerial discretion to pursue the public interest even when that undoubtedly sacrifices profits. It is to those affirmative gains that I turn next.

Why Some Managerial Discretion to Sacrifice Profits in the Public Interest Is Affirmatively Desirable and Efficient

Shareholder profit maximization leaves out much that is relevant to overall social efficiency. To at least some extent, shareholders value nonfinancial aspects of corporate activities, such as whether those activities further the shareholders' social and moral views. Thus, maximizing shareholder welfare is not the same thing as maximizing shareholder profits. Further, limiting the inquiry to shareholder welfare leaves out any harm a corporation might inflict on interests outside the corporation, including on other corporations. Considering these other factors reveals that managerial discretion to sacrifice corporate profits in the public interest is not just inevitable, but affirmatively desirable.

Why Shareholder Welfare Maximization Affirmatively Justifies Sacrificing Corporate Profits in the Public Interest Whenever Managers Are Acting as Loyal Agents for Most Shareholders. Consider first the reality that some shareholders derive nonfinancial benefit from having corporate activities further their social and moral views, or would suffer social or moral sanctions from corporate violations of social or moral norms. This is certainly true for controlling shareholders such as Henry Ford or Philip Wrigley, who are heavily involved in their firm's management. It is also true for large, but not controlling, investors such as Warren Buffet, who "says explicitly that he is willing to sacrifice the financial interests of shareholders in favor of 'social' considerations."[85] It is even true for

85. *See* Henry Hu, *Buffett, Corporate Objectives, and the Nature of Sheep,* 19 Cardozo L. Rev. 379, 391–92 (1997).

many shareholders with smaller investments in public corporations. An increasing number of investors now put their money in funds that commit to avoiding investments in corporations that create environmental harms; make tobacco, alcohol, or weapons; or engage in some other activity that conflicts with various conceptions of the public interest. In 1997, $1.185 trillion, or 9 percent of all investments, was managed using some form of social screen.[86] By 1999, the figure was $1.5 trillion.[87] Likewise, investors increasingly are government pension funds, unions, and university endowments, which in part often have nonfinancial agendas.[88] For such shareholders, their welfare reflects a combination of their financial returns and their social or moral satisfaction with corporate activities.

An enforceable duty to profit-maximize thus would decrease shareholder welfare whenever the harm that shareholders suffer from decreased social or moral satisfaction with corporate activities exceeds the gain they derive from increased profits. Suppose, for example, that installing lights in Wrigley Field would increase the Cubs' profits by $1 million but cause Mr. Wrigley to suffer a disutility that he values at $2 million. If Mr. Wrigley were a sole proprietor, he could take this into account and refuse to install lights even though it would maximize his profits. But if using the corporate form to run the Cubs required him to profit-maximize, then the Cubs would have to install lights. As 80 percent shareholder, Mr. Wrigley would gain profits of $800,000 but suffer a net welfare loss of $1.2 million. If the 20 percent shareholders do not share his love for daytime baseball, they would gain $200,000. The net result is that enforcing a duty to profit-maximize would inefficiently result in a loss of shareholder welfare worth $1 million.

The issue would be the same for our clear-cutting corporation if enough shareholders suffer a disutility from social or moral sanctions on clear-cutting to exceed any profit gain. A duty to profit-maximize in that case would require the corporation to engage in conduct that inflicts social and moral sanctions on shareholders even when those shareholders would prefer otherwise.

To the extent managers are acting as loyal agents for the majority of shareholders, managerial decisions to sacrifice profits in the public interest will by definition increase the welfare of most shareholders. But most is not all, which raises the commonly made objection that those running a corporation should not be able to "tax" dissenting shareholders to further public-interest objectives that the dissenting shareholders have not chosen and may not share. As Dean Clark nicely put the argument, "It is morally good to be generous, but please be

86. *See* Cynthia A. Williams, *The Securities and Exchange Commission and Corporate Social Transparency*, 112 HARV. L. REV. 1197, 1287–88 (1999).

87. Michael S. Knoll, *Ethical Screening in Modern Financial Markets*, 57 BUS. LAW. 681, 681 (2002).

88. *See* Douglas M. Branson, *Corporate Social Responsibility Redux*, 76 TUL. L. REV. 1207, 1219 (2002).

generous with your own money, not that of other persons."[89] Although majority shareholders who choose to sacrifice corporate profits would be taxing themselves as well, the objection remains that they also would be forcing the minority to fund a percentage of public-interest objectives that the majority has chosen. In the Wrigley example, the objection would be that it is fine for Mr. Wrigley to increase his utility by spending $1 million of his own money, but not by getting 20 percent of that $1 million from nonconsenting shareholders.

However, this "tax" argument founders on closer analysis. To begin with, managers have no choice but to make an operational choice that will disappoint some shareholders. The corporation cannot be operated in different ways for different shareholders. Either lights will be installed, harming the 80 percent shareholder, or they won't be installed, harming the other 20 percent. Either clear-cutting will begin, harming the shareholders who morally oppose it, or it won't, harming the shareholders who care only about their financial returns. One may think it unfair to allow the majority shareholders to force dissenting shareholders to forgo profits to further the causes of the majority shareholders. But wouldn't it be even more unfair to give any dissenting shareholder the power to force the majority shareholders to operate the corporation in a way they feel is wrong or immoral, especially when that exposes them to social and moral sanctions or otherwise decreases overall shareholder welfare?

One implication of this is that, even if one thought maximizing profits should be an enforceable duty, enforcing it with injunctive relief would be undesirable. One might, however, argue that the proper solution would be to allow controlling shareholders to make the operational decision that sacrificed profits to further some public-interest objective by denying any claim for injunctive relief, but still oblige them to reimburse the corporation for any sacrificed profits by recognizing an action for damages. This would both avoid any "tax" on dissenting shareholders and make sure that the controlling parties really enjoy a welfare gain that exceeds the lost profits. Mr. Wrigley could keep lights out of Wrigley Field, but he'd have to pay $1 million to the corporation owning the Cubs to compensate it for the lost profits. The dissenting shareholders would be in the same financial position as if lights had been installed and thus could not claim to have been taxed. While Mr. Wrigley's welfare gain would no longer be the $1.2 million he would enjoy without an enforceable duty, it would still be $1 million compared to his situation if lights had been installed.

But even limited to a claim for damages, this "tax" argument remains flawed because it implicitly assumes as a baseline the very issue being debated. Terming a decision to sacrifice profits a "tax" or a use of money belonging to "other persons" implicitly assumes a baseline of pure profit maximization to which shareholders are entitled, so that any deviation equals a "tax." If we instead assume that corporate managers can pursue some unprofitable social activities, then the question becomes why dissenting shareholders should be able to "tax" controlling parties for the exercise of their right to pursue nonprofit objectives. And if

---

89. Clark, *supra* note 1, at 603, 679; *see also* Milton Friedman, *supra* note 1, at 33.

the money never "belonged" to the dissenting shareholders in the first place, then they cannot be said to have been "taxed" out of it. For example, if Mr. Wrigley had the right not to install lights, why should the dissenting shareholders be able to "tax" him $200,000 for exercising it? Because the "tax" argument depends on the existence of a baseline, it would be circular to employ it as an argument about what that baseline should be.

One possible way to set the baseline would be by shareholder expectations. If shareholders buy into corporations knowing that they are run by managers and controlling shareholders who can temper profit maximization, then shareholders will have bought in at lower stock prices that reflected that fact and can claim no tax or injury when that tempering occurs. On the other hand, if shareholders buy their shares expecting pure profit seeking, then they will have bought at prices that reflected that expectation and thus will suffer a loss if profits are sacrificed. Nor can they avoid the economic loss that results when a corporation embarks on a course of sacrificing profits by just selling their shares because the expected decline in future earnings will be capitalized into the market price at which they can sell their shares. Unfortunately, solid empirical evidence on shareholder expectations that explores the relevant nuances appears lacking. But it seems clear that while shareholders expect profits and do not regard stock investments as tantamount to charitable contributions, they also do not expect unabashed profit seeking untempered by any sense of social responsibility. Expectations of pure profit maximization seem particularly unlikely in light of the laws noted above that expressly authorize charitable contributions and countenance other forms of public-spirited activities.

As this last point reveals, the deeper problem is that the expectations argument, like the tax argument, is ultimately circular. Shareholder expectations are likely to reflect the governing legal regime. If that regime mandates pure profit maximization, shareholders will expect pure profit maximization. If that regime allows corporations to temper profit maximization, shareholders will expect tempering. Whatever shareholder expectations happen to be at the moment is not that important;[90] the ultimate policy question is what expectations we want shareholders to have.

A better argument for a profit-maximization duty enforceable by money damages is that requiring controlling shareholders to pay the corporation for any lost profits would force them to internalize the "externality" that their decision imposed on other shareholders, and thus make sure that the profit-sacrificing conduct really did increase total shareholder welfare. Suppose, for example, that Mr. Wrigley's utility benefit from daytime Cubs baseball was only $900,000. Then without any duty he would not install any lights because this utility benefit would exceed the $800,000 in profits he would lose given his 80 percent share

90. It may raise important transitional problems, though typically the degree of reliance parties place on any status quo will be more efficient if the relying parties bear the risk that the status quo might change. *See* Louis Kaplow, *An Economic Analysis of Legal Transitions*, 99 HARV. L. REV. 509 (1986).

of the corporation's $1 million sacrifice. But his net welfare gain of $100,000 here would be less than the $200,000 the other shareholders would lose. In contrast, with a duty enforceable by money damages, he would refrain from installing lights only if his utility benefit from daytime baseball were really greater than the $1 million in lost corporate profits, which is the same as when his net welfare gain from not installing lights exceeds the $200,000 in harm to dissenting shareholders who care only about profits.

However, this "externality" argument for setting the baseline at profit-maximization runs into three other problems. *First*, this argument does not work if other shareholders share the public-interest view of the controlling shareholder and thus experience uncompensated *positive* externalities from his profit-sacrificing conduct. Suppose, for example, we added to the hypothetical in the last paragraph the fact that the 20 percent of shares not owned by Mr. Wrigley are held equally by 200 shareholders, 150 of whom actually share Mr. Wrigley's view about daytime baseball and would each experience a utility loss worth $2,000 if lights were installed, which exceeds the $1,000 in additional profits they would earn. The other 50 shareholders care only about profits and threaten to sue Mr. Wrigley. Then Mr. Wrigley would install lights because his $100,000 welfare gain from daytime Cubs baseball is less than the $200,000 in liability he would have to pay. Yet installing lights would decrease shareholder welfare by a total of $1.2 million to earn an additional $1 million, which is inefficient.

One might imagine trying to avoid this problem by making damages payable not to the corporation for all lost profits but only to dissenting shareholders for their proportionate share of lost profits. But then each of the 150 shareholders who actually agree with Mr. Wrigley would have an incentive to dissent, reasoning that her individual decision to dissent would gain her $1,000 in damages but have little impact on whether lights were installed. If she thought all the other 149 shareholders wouldn't dissent, then she would figure that lights wouldn't be installed no matter what she did, so she might as well dissent and get $1,000 on top of her $2,000 utility benefit. If she thought all the other 149 shareholders would dissent, then she would figure that Mr. Wrigley would decide to install the lights no matter what she did, so she might as well reduce her $2,000 utility loss by dissenting and taking the $1,000. In short, no matter what she thought the other like-minded shareholders would do, collective-action problems would give each incentives to dissent, even though the result of all of them dissenting would end up being harmful to their welfare. The same result would follow if damages were made payable to the corporation not only by the controlling shareholder, but also by all the other shareholders who indicated they agreed with his profit-sacrificing decision. Collective-action problems then would cause each shareholder to indicate disagreement even when he actually agreed because his individual decision would have little impact on what corporate conduct occurred but would definitely avoid personal liability.

Second, as the Coase Theorem has taught us, one can always flip any externality argument around, so that the characterization of something as an "externality" also has baseline problems.[91] Suppose, for example, we assumed that all the shareholders who held the 20 percent of stock not owned by Mr. Wrigley were pure profit maximizers. If the law allowed Mr. Wrigley not to install lights, and he really would get only a net $100,000 benefit from doing so, then the other shareholders should be willing to pay him something between $100,000 and $200,000 to agree to install lights in Wrigley Field. Forcing the other shareholders to do so would make sure that they internalized the "externality" that profit-maximizing light installation would impose on Mr. Wrigley's welfare.

One need not rely on having the other shareholders organize themselves to make a payment to Mr. Wrigley, which would present collective-action problems. If the $200,000 harm in lost profits experienced by other shareholders exceeds Mr. Wrigley's net utility benefit, then Mr. Wrigley would have incentives to do a freeze-out merger that merges the existing firm into a new corporation owned solely by him, and have that new corporation commit to install lights.[92] If he wanted to remain an 80 percent shareholder, he could do so by selling 20 percent of the shares in the new corporation for a stock price that should be $200,000 higher than the price for 20 percent of the old Cubs corporation. Or he could simply sell 100 percent of the shares in the new corporation to new profit-maximizing managers who should be willing to pay $1 million more than the old stock price, which by hypothesis would exceed his utility benefit from control. A controlling shareholder such as Wrigley would have an incentive to pursue one of those tactics whenever the monetary loss from his profit-sacrificing activity exceeded the utility benefit he derived from it.

In short, as the Coase Theorem further teaches, the efficient result will occur regardless of the initial legal entitlement as long as transaction costs are zero, so the real issue is which initial entitlement minimizes transaction costs. And here transaction cost considerations would seem to strongly favor giving the initial entitlement to the controlling shareholder. As the Wrigley examples have shown, if the controlling shareholder has the discretion to sacrifice profits, then opting out of that discretion whenever it decreases total shareholder welfare is a mere matter of paperwork within his control, which imposes little transaction cost. In contrast, the transaction costs of enforcing a duty to profit-maximize are extremely high because litigation is: (1) highly expensive and contentious; (2) risky for dissenting shareholders to fund given that it might lose; and (3) unlikely to be effective because courts have such great difficulty figuring out what maximizes profits that they will either frequently condemn profit-increasing activities by mistake or give a business judgment deference that makes enforcement impracticable.[93]

91. R.H. Coase, *The Problem of Social Cost*, 3 J.L. & Econ. 1, 2–8 (1960).

92. The commitment could be made by a charter provision, by contracting for light installation, or by making sunk investments in light installation.

93. This is true whether the duty is enforced via a claim for damages or injunctive relief.

Further, where other dispersed shareholders share the public-interest views of the controlling shareholder, collective-action problems are likely to impose an insuperable transaction cost on getting them to contribute to help the controlling shareholder pay damages for profit-sacrificing activity, even when that activity actually enhances the welfare of those shareholders. Nor, unlike in the reverse case, could one overcome such collective-action problems by merging the firm into a new corporation with a charter provision opting out of the initial entitlement, here by allowing the profit-sacrificing activity. The reason is that if the initial entitlement is profit maximization, then all the old shareholders would have to be bought out at a high price that reflected profit-maximizing conduct, but even the old shareholders who agreed with the controlling shareholder would not be willing to pay a price for the new corporation that included the utility benefit they derived from profit-sacrificing conduct because they would figure they would get that utility benefit whether or not they invested in the new corporation. For example, the 150 shareholders who hypothetically agreed with Mr. Wrigley would in a first step merger get a price reflecting the $1,000 profits that light installation would produce, but they would not pay an additional $2,000 for new shares in a corporation with a provision prohibiting light installation because they would figure that the corporation wouldn't install lights no matter what they did.

Thus the solution most likely to minimize transactions costs and maximize shareholder utility would be to give the controlling shareholder the right to make operational decisions that sacrificed corporate profits to further his conception of the public interest. As long as shareholders who cared about only profits bought their shares understanding this was the rule, they would not suffer any loss, for the price they paid for their shares would reflect this rule. But the controlling shareholder and others who shared his public-interest views would gain, as would overall shareholder welfare.

Third, the argument that a profit-maximization duty enforceable by money damages makes the controlling parties internalize the externality their profit-sacrificing decisions impose on dissenting shareholders does not work when there is no controlling shareholder, but instead corporate managers who are acting on behalf of the public-interest views of dispersed public shareholders. The liability for sacrificed profits would then be imposed on the managers, not the like-minded shareholders, and thus give managers incentives to have the corporation engage in profit-maximizing activity even when the conflict with the public-interest views of shareholders caused a net loss of shareholder welfare.

Again, one could imagine trying to change this by having shareholders vote on whether to engage in such profit-sacrificing behavior, and making shareholders liable for a share of the lost profits if they voted for such behavior or eligible to receive damages only if they dissented from it. But in a public corporation with dispersed shareholders, such a rule would raise the same collective-action problems as noted above. Each shareholder would dissent even when they agreed with management because dissenting would get them a monetary gain without having any significant impact on whether an operational decision collectively is made that decreases shareholder welfare given their public-interest views.

These collective-action problems indicate that, even if we did call the profit-sacrificing conduct a "tax" on shareholders, such a tax might be justified on the ground that coercive financing can be necessary to overcome collective-action problems among shareholders. Without such coercive financing, shareholders who in their heart of hearts would rather sacrifice profits to advance the public interest would have an incentive to dissent, hoping that the majority will still further the public-spirited activity, but without their contribution. They will, in other words, free ride without coercive financing. Their position is precisely analogous to the citizen who would vote for building a dam with tax money coercively taken from everyone, but would not voluntarily make an individual contribution if someone went door-to-door seeking contributions to build the dam.

One might object that most shareholders must instead prefer profit maximization because the lion's share of investors do *not* invest in socially conscious funds. But the same collective-action problems that were just noted also mean that individuals would have little incentive to invest in funds that sacrificed additional profits to further their public-interest views. After all, their individual decision to invest in such a fund would definitely reap them a monetary loss but have little impact on whether such funds were generally successful in changing corporate conduct. Thus even if investors do have public-interest views, they will have little incentive to act on them when making investment decisions among funds or corporations. Given this, it is remarkable that many people do invest in socially responsible funds.

Although collective-action problems mean that shareholder investment decisions should reflect very little of the utility that shareholders derive from socially responsible corporate activities, this does not at all mean that shareholders do not in fact derive significant utility from corporate conduct that sacrifices profits to further the public interest. Political polls and the behavior of shareholders as voters in the political process suggest they are strongly influenced by public-interest views. Further, one survey found that 97 percent of corporate shareholders agreed (75 percent strongly) that managers should consider other constituency interests, and 88.5 percent agreed that managers considering a plant move that would be profitable to shareholders "should weigh the effect the move would have on its employees, customers, suppliers and people in the community it presently is in before deciding to move."[94] This should not be too surprising, for the social and moral norms likely to guide managerial discretion generally are broad-based enough that they probably will be shared by most shareholders. This is especially likely because, in their personal capacities, corporate shareholders are likely to be among the noncorporate parties who might be

94. *See* Larry D. Soderquist & Robert P. Vecchio, *Reconciling Shareholders' Rights and Corporate Responsibility: New Guidelines for Management,* 1978 DUKE L.J. 819, 841 tbl.3 (1978).

adversely affected by corporate activities; they as a group will encompass the employees, bondholders, community members, or citizens harmed by environmental pollution. Thus, although for specific corporations shareholders may not be in the harmed group, they are likely to benefit from any social or moral norms that generally prevent corporations from unduly harming others.[95]

If managers are *not* acting as loyal agents for most shareholders, then their exercises of profit-sacrificing discretion may be harmful to shareholder welfare. Managers may further conceptions of the public interest that their shareholders find disagreeable, or weigh any public-interest considerations more heavily than the utility that shareholders would derive from them. But if that is the objection, it indicates that managers should be free to engage in profit-sacrificing conduct that a majority of shareholders has approved. Likewise, even absent such an affirmative vote, the fact is that managers are elected by a majority of shareholders. Thus, absent evidence to the contrary, one might presume that managerial tradeoffs between profits and public-interest considerations likely reflect the views of most shareholders.

For example, suppose the Cubs stock was held completely by dispersed shareholders, and corporate managers refused to install lights even though their decision sacrificed profits. If a majority of the dispersed shares were held by shareholders who derived no utility from maintaining daytime baseball, then one would think they would elect new managers who would install lights. If a majority of shareholders do derive utility from daytime baseball that exceeds the lost profits, then they should elect managers who would not install lights. In making such a voting decision, the dispersed majority shareholders would not face the same collective-action problems that plague them in the other situations noted above because they would not be making the sort of decision that separates their individual gain from the collective decision. If they vote for managers who will install lights, they get the combination of increased profits and lost utility for those who like daytime baseball, and if they vote for managers who won't, they get the inverse combination. Their individual voting decision may have little impact on the ultimate corporate activity, but it also will have equivalently little impact on whether they get the resulting profits, so their voting decision does not suffer from the sort of disjunctions noted above.

Generally, total shareholder welfare will be maximized by making the decision that increases welfare for most of the shareholders. Still, in some cases, the concern remains that most shareholders might derive a welfare gain from profit-sacrificing corporate activities that exceeds their share of lost profits, but does not exceed total lost profits when one includes dissenting shareholders who do not share their public-interest views. For example, if 80 percent of the Cubs' shares are held by dispersed shareholders who derive a utility benefit from

---

95. *See* JAMES P. HAWLEY & ANDREW T. WILLIAMS, THE RISE OF FIDUCIARY CAPITALISM: HOW INSTITUTIONAL SHAREHOLDERS CAN MAKE CORPORATE AMERICA MORE DEMOCRATIC 1–30 (U. Penn Press, 2000) (arguing that a fully diversified "universal shareholder" has holdings of different corporations, bonds as well as stock, and human capital as well as financial capital, and thus would not want a corporation to harm other corporations, bondholders, or employees).

$900,000 from daytime baseball, but the total lost profits are $1,000,000, then a natural concern is that managers acting on behalf of the 80 percent would fail to install lights because that increases the welfare of those shareholders even though it decreases total shareholder welfare. Still, over time, one would think that the dissenting shareholders would sell their shares to shareholders who do share the public-interest views of the majority shareholders and thus derive extra utility from it. The dissenting shareholders would receive a price that reflected the profit-sacrificing conduct, but if that is the expected rule, then that is also the price at which they would have purchased. The corporation would be left with a final set of shareholders who shared the majority shareholder view about daytime baseball, and either would derive enough utility from that to exceed their lost profits (in which case shareholder welfare is enhanced by continuing daytime baseball) or wouldn't (in which case they would elect managers who would install lights).[96]

True, shareholders can exercise such oversight only if they are accurately informed about the profit-sacrificing conduct of managers. But this is yet another argument for allowing patent exercises of profit-sacrificing discretion. If managers can sacrifice profits only surreptitiously, by making bogus claims that the conduct really enhances profits, then shareholders will have difficulty becoming informed about what actually is going on. Shareholders would have to be sufficiently informed about corporate activities to second-guess managerial assertions about what maximizes profits, ascertain how much in profits is being sacrificed, and determine what public-interest justifications might be furthered. In contrast, if managers explicitly sacrifice corporate profits in the public interest, then shareholders will be alerted to what is going on and thus made aware of the need to focus on the issue, will be informed about the profit sacrifice and public-interest justification at issue, and therefore will be better placed to judge both whether they agree with managers about the public-interest justification and whether it merits the lost profits.

Patent profit sacrificing by managers thus produces more informed decision-making by shareholders, and is more likely to advance shareholder welfare. Latent profit sacrificing increases information costs for shareholders, and thus is less likely to advance shareholder welfare and more likely to increase agency slack. This is one important reason for moving beyond the latent profit-sacrificing discretion conferred by the business judgment rule to the patent discretion that the law also recognizes.

In short, this section shows that shareholder welfare would be maximized by a rule that allowed controlling shareholders to sacrifice profits in the public

96. Another possibility is that the 20 percent of shareholders who are purely profit maximizers would pay the other 80 percent of shareholders $100,000 to $200,000 to install lights. But unlike in the case of a controlling shareholder, it is difficult to see how they could enter into an enforceable transaction that allocated that payment to just the 80 percent of shareholders who agreed with managers without raising collective-action problems about the incentives of shareholders to accurately identify which group they fall into.

interest and allowed managers with dispersed shareholders to do so when they have majority shareholder support. The latter is clear when the shareholders have in fact voted for the specific operational decision in question. It also is likely to be true when managers who have been elected by shareholders are making profit-sacrificing decisions that have not been rejected explicitly by most shareholders. Where managers are acting as loyal agents representing the views of most shareholders, allowing managers to sacrifice profits in the public interest will maximize shareholder welfare. Those dissenting shareholders who care only about profits will receive just the economic return that they should have expected, given the legal rule allowing such discretion. The other shareholders who get net utility from having profit-sacrificing corporate activities further public-interest activities will enjoy greater welfare and have incentives to vote to alter the corporate decision whenever that ceases to be true.

Why Social Efficiency Affirmatively Justifies Giving Managers Discretion to Sacrifice Corporate Profits in the Public Interest Even When They Are Exercising Agency Slack. Sometimes managerial decisions to sacrifice profits in publicly held corporations will not maximize shareholder welfare. That seems clearly true when managers sacrifice profits over the explicit objection of most shareholders. More generally, it is true when managers are exercising the agency slack left to them because shareholder monitoring is imperfect. Even though most managerial decisions should conform to shareholder welfare, considerable agency slack will be left when shareholders are dispersed because collective-action problems undermine shareholder incentives to become informed before voting or even to exert the effort to read and assess any information disclosed to them. Each shareholder will know that if she expends the cost of making a better-informed vote, her vote will have little impact on the outcome, so she might as well not expend that cost and remain uninformed.

Within this zone of agency slack, managers might engage in more or less profit sacrificing than shareholders would want, or further public-interest views that conflict with those of shareholders. Does this justify creating an enforceable duty to profit-maximize? No. Such managerial deviations from shareholder views are affirmatively likely to improve corporate conduct because shareholder insulation and collective-action problems will make shareholders underresponsive to social and moral sanctions. And imposing an enforceable duty to profit would make corporate behavior even worse vis-à-vis third parties. Moreover, it likely would harm shareholders by interfering with both business judgment deference and those exercises of profit-sacrificing discretion that benefit most shareholders.

Why the Corporate Structure Means that Managers Improve Corporate Conduct When They Exercise Their Agency Slack to Respond to Social and Moral Sanctions. As discussed in the section on the social regulation of noncorporate conduct, optimizing conduct has always required supplementing legal and economic sanctions with social and moral processes. In a noncorporate business such as a sole proprietorship or general partnership, the owners play an important role in managing the business and thus become subject to a host of social or moral processes that guide their behavior in non–profit-maximizing ways. In part, these processes work by subjecting business owners to the usual set of social and moral sanctions that attend antisocial behavior even when it is legal. In part, however, these social or moral processes work by creating a greater awareness that comes from confrontation with problems and the results of one's actions. The manager who sees his workers suffer under a poor working environment will, if at all motivated by a concern for others, be more likely to improve those working conditions. Finally, these social and moral processes in part involve a creation of private values to which economic theory cannot speak because it takes people's values as givens. People can be socially molded to derive personal gratification from doing good. For example, social processes can make materialists into philanthropists, creating values that make the philanthropists feel good when they donate money to worthy causes. This results in a state of the world better than any possible without the creation of those values. Unlike with taxation, the philanthropists' incentives to create wealth are not diminished because they feel as much pleasure (with their new values) from donating the money as they would have felt from buying a Porsche. And yet the same sort of redistribution is accomplished that otherwise would have required taxation. Thus it is not surprising that noncorporate businesses have always felt some social responsibility to contribute to the community—sometimes informed by an enlightened view of their long-term financial interest, but often based on nonfinancial grounds.

On these social and moral dimensions, corporations historically have been viewed with great suspicion. The "old maxim of the common law" was that "corporations have no souls."[97] This was more than a minor concern: the "soulless" nature of corporation was a source of great opposition to chartering corporations at all in the nineteenth century:

> The word "soulless" constantly recurs in debates over corporations. Everyone knew that corporations were really run by human beings. Yet the metaphor was not entirely pointless. Corporations did not die, and had no ultimate limit to their size. There were no natural bounds to their life or to their greed. Corporations, it was feared, could concentrate the worst urges of whole groups of men; the economic power of a corporation would not be tempered by the mentality of any one person, or by considerations of family or morality.[98]

But why should corporations, which after all are owned and run by humans, be feared more than ordinary businesses? The answer that they are large and never die hardly seems satisfactory, both because that can be true of noncorpo-

97. Lawrence Friedman, A History of American Law 448. (1st ed. 1973).
98. *Id.* at 171–72.

rate business enterprises and because one would think that humane considerations would nonetheless tug at the human managers running even a huge and immortal organization.

Although not well expressed at the time, a better answer lies in the corporate structure, which raises two important obstacles for a regime that relies in part on social and moral processes to guide behavior. First, the corporate structure largely insulates shareholders from the ordinary social and moral sanctions that a sole proprietor would feel, especially in the large, publicly held corporation that raises the concern we are now addressing about managers exploiting agency slack. Shareholders are less likely to come into contact with those who might want to impose social sanctions for the business's illegal activities and will be harder to identify as being connected to the corporation at all. Moral sanctions are not susceptible to those problems, but raise different ones because moral sanctions require knowing just what the corporation is doing, and shareholders ordinarily are blissfully unaware about the details of operational decisions. Certainly they lack the sort of detailed and vivid information about how corporate operations may impact the public interest that is necessary to create strong feelings of guilt. A shareholder will not feel much moral guilt about his corporation's clear-cutting, for example, if he isn't sure whether the corporation really is doing it, how bad its environmental effects are, or whether these effects are offset by favorable employment effects. Even if these obstacles could be overcome, shareholders are less likely to be deemed or feel responsible because each is one of many shareholders. This diffused responsibility should further insulate shareholders from social or moral sanctions. Separating ownership from management of corporate operations also means the owner–shareholders will not participate in the sort of social and moral processes that give businessmen affirmative desires to behave in socially desirable ways when the law and profit motives are insufficient to do so.

So uninformed and shielded, shareholders in publicly held corporations will suffer much lower social or moral sanctions from undesirable corporate conduct than will a sole proprietor engaged in the same business conduct. Given the inevitable underinclusion of even optimal legal regulation, these social and moral sanctions are necessary to optimize behavior even outside the bounds of illegality. Thus a corporation whose managers always acted to maximize shareholder welfare likely would engage in more socially undesirable behavior than would a sole proprietor because the social and moral sanctions on those shareholders are so much lower. Instead, we should expect corporate shareholders to be more relentless than other business owners in pressing managers for unabashed profit seeking untempered by social consequences because they don't have the knowledge to feel moral guilt or the social exposure to feel social sanctions. A corporation run by managers perfectly accountable to shareholders would be "soulless" because the corporate structure insulates shareholders from the social and moral processes that give us "soul."

Second, dispersed public shareholders have collective-action problems that make it difficult for them to act on any social or moral impulses they do feel. This is certainly true when making investment decisions. Each shareholder deciding whether to buy or sell stock in a particular public corporation will know that her investment decision definitely will determine whether she gets a share of the associated profits but will have little impact on whether the corporation engages in the conduct that offends her social and moral sensibilities. These collective-action problems mean that shareholder investment decisions will not tend to drive down the stock market prices of corporations that violate social and moral norms even to the extent that shareholders do care about those norms despite their insulation. To the contrary, the investment decisions of even caring and informed shareholders will tend to drive down the stock prices of corporations that sacrifice profits to comply with social and moral norms that investors themselves hold.

Likewise, because each individual shareholder has little impact on who wins any shareholder vote, each will also have little incentive to expend energy on collecting or even reading information about operational decisions before they vote. Even if a shareholder cares deeply about the environment and receives information about a corporation's clear-cutting, she won't have incentives to spend time reading it, let alone checking it against other sources of information to determine if it is accurate in its claims. For if she spends all that time to make her vote a more informed one, she knows that her single vote remains highly unlikely to alter the outcome.

The historical response to such fears about corporate soullessness rested largely on assurances that society could trust in the souls of the humans who managed them. To the extent this response was persuasive, I think what it meant was that managers would be subject to social and moral sanctions, pressures, and processes that would tend to counteract their accountability to shareholders. People will protest outside managers' offices, letters will flow into their mailboxes, and the applause from good corporate conduct will ring in their ears. And managers will know what the corporation is doing and see its effects sufficiently to experience moral guilt for causing any ill effects that violate moral norms. Managerial responsiveness to social and moral sanctions should thus compensate for shareholder pressure to ignore those social and moral sanctions. This is consistent with the fact that, although shareholder proposals on social responsibility are often made, they usually lose overwhelmingly with shareholders, and normally are more successful in persuading management than shareholders.

But this historical response could make sense only if managers have some legal discretion to use their agency slack to sacrifice corporate profits in the public interest even when shareholders are indicating otherwise in their votes or investment decisions. By eliminating that discretion, a legal duty to profit-maximize would take away the human element that helps justify allowing the use of the corporate form at all.

Indeed, creating an enforceable duty to profit-maximize would worsen the problems created by corporate structure in two ways. First, a corporation whose behavior was governed solely by an enforceable duty to profit-maximize would be forced to engage in the sort of suboptimal conduct we would get with *zero* social and moral sanctions. This will worsen corporate conduct even more than what would result from mere accountability to shareholders who are insulated from those social and moral sanctions.

Second, because the duty could be enforced by any single shareholder in a derivative action, it would dictate corporate governance by the lowest common moral denominator: that is, by whichever shareholder cares least about social and moral sanctions. Even if the average shareholder would feel the same social and moral sanctions as the average sole proprietor, such a duty would leave corporate behavior dictated by the subaverage shareholder who feels lower social and moral sanctions, and thus make corporate behavior worse than the average behavior of a sole proprietor. Given the actual insulation of average shareholders, such a duty would make corporate behavior even worse than the average wishes of shareholders who already are underresponsive to social and moral sanctions. This would thus result in even greater underresponsiveness to social and moral sanctions than accountability to shareholders alone could produce. Such a duty to profit-maximize would allow any minority shareholder to sue all the other shareholders into ignoring their sense of social responsibility—thus enforcing the very soullessness for which corporations historically have been feared.

Proponents of a duty to profit-maximize have ignored these issues because they assumed away any role for social and moral sanctions when they assumed that any legitimate public-interest objectives could be embodied in legal regulation. They argued that business operations could be regulated (by laws applicable to corporate and noncorporate businesses) to fully protect or compensate nonshareholder groups who might be injured by those operations, that the corporate profits that would be increased by a duty to profit-maximize could be taxed to fund public goods or further goals of equitable wealth distribution, or that some combination of strategies could be employed to ensure that the end result is Pareto optimal.[99] Duty proponents further argued that even when general regulation was insufficient, other stakeholders could also protect themselves with legal contracts with the corporation, relying on judges to fill gaps in those contracts to fine-tune that protection when unforeseen events arise.[100] Because their assumptions meant that the public interest was or could be fully taken into account by the law, duty proponents could then argue that legal profit-maximizing corporate conduct not only would increase national wealth and encourage

---

99. Clark, *supra* note 1, at 20–21, 680; Frank H. Easterbrook & Daniel R. Fischel, The Economic Structure of Corporate Law 37–39 (1991); Hansmann & Kraakman, *supra* note 1, at 441–42; Macey, *supra* note 1, at 42–43.

100. *See* Hansmann & Kraakman, *supra* note 1, at 441; Macey, *supra* note 1, at 40–41.

shareholder investment, but also would be socially desirable.[101] Others, who do not quite advocate profit maximization but favor relatively narrow profit-sacrificing discretion, likewise have relied on a similar premise that legislative action and inaction reflect a long-run political consensus about what is desirable.[102]

But as detailed in the section on the social regulation of noncorporate conduct, this belief in the perfection or even perfectability of law is misplaced. Instead, even the most efficient and socially optimal legal rules will fail to cover much undesirable conduct. Thus corporate conformity with the law does not suffice to render corporate conduct socially desirable. Nor can we be sure that corporate profit making within legal limits will be efficient from the societal perspective; because of the inevitable imperfections of law, it may impose harms that exceed the benefits of the extra profits.

In addition, often the types or magnitudes of harm that corporations inflict on nonshareholder groups change before the government has time to act, especially given the usual lag time for governmental action.[103] This cannot be corrected by simply making governments act faster because there is always a balance between speed and spending the time necessary to secure the knowledge, deliberation, or social consensus that gives us some assurance that the governmental action is in the public interest. Even ignoring delays in timing, a separate problem is that it takes great efforts and often significant resources to secure governmental action, thus frequently making it more efficient (from the perspectives both of the affected interests and of society) to lobby corporations directly with social and moral pressure. The effort to legally define and enforce public-interest objectives, in other words, will often rationally be avoided by society and the participants because the net benefits of obtaining legal definition and enforcement (compared to relying on social and moral sanctions) will not be worth the costs.

All these problems are further complicated by the fact that many corporations do business in numerous nations with varying legal standards. For example, before 1924, slavery was legal in the Sudan and not yet prohibited by international law.[104] Even if engaging in slavery in the Sudan in 1920 would have maximized profits, presumably no court would have held that a U.S. corporation was then obligated to engage in Sudanese slavery when doing so was so contrary to social and moral norms held dear here. But then that would mean that the duty

---

101. CLARK, *supra* note 1, at 20–21, 679–80, 692, 702. A related argument is that nonprofit corporations exist to pursue public-interest goals. This is true, but it provides no reason not to have mixed-purpose organizations. Nor does it justify preventing business corporations from running their operations in a manner that best advances the public interest.

102. *See* David L. Engel, *An Approach to Corporate Social Responsibility*, 32 STAN. L. REV. 1, 2, 34–37 (1979).

103. *See* C. STONE, WHERE THE LAW ENDS 94–96 (1975).

104. *See* Report of the International Eminent Persons Group, *Slavery, Abduction and Forced Servitude in Sudan* 19–24 (May 22, 2002), obtained at http://www.state.gov/p/af/rls/rpt/10445.htm#legal.

did not really require all legal profit-maximizing activities, but picked among them based on the strength of the social or moral norm against it. Alternatively, the courts could require a U.S. corporation to comply with U.S. law even when operating abroad. But if applied to all laws, that would subject U.S. corporations to disadvantageous regulations in foreign nations that those nations did not even want, such as perhaps tough environmental regulations that make sense in the United States but not an undeveloped nation. Slavery in the Sudan is admittedly an extreme case, but the general point remains valid: variations in legal regulation among different nations inevitably will leave legal gaps requiring supplementation by social and moral sanctions that operate internationally.

In short, legal regulation is an important but insufficient means of policing behavior, be it the behavior of individuals, noncorporate businesses, or corporations. Accordingly, Dean Clark's proposal—that if current law fails to capture public-interest goals that corporations can further, then we should just redouble our efforts to define public-policy objectives and determine when it is wise to contract out implementation of those objectives to profit or nonprofit corporations[105]—is fine as far as it goes, but incomplete. It fails to face up to the fact that no matter how energetic our efforts, any lawmaking process will have defects, any legal definition will be imprecise, and the costs of legal definition and enforcement will often exceed the benefits. Because of these inherent limits with legal regulation of behavior, social and moral sanctions will always play an important supplemental role in maximizing the likelihood of desirable behavior.

It should not be surprising if, as Dean Clark asserts, lawyers and economists commonly assume that the corporations need only profit-maximize within the law to assure that their behavior is socially desirable,[106] for that position reflects an exaggerated view of the importance of both fields: lawyers who overestimate the influence of the law, and economists who overestimate the importance of financially self-interested behavior. Nor should it be at all surprising that those actually subject to the social and moral processes that play such an important role in real life—that is, corporate managers—persist in having a far different view of their role. Groups that represent corporate management, such as Business Roundtable, "have denied that profit-maximization should be the basic criterion by which managements should be judged."[107] Surveys indicate that most managers believe that they must weigh shareholder interests against those of other stakeholders.[108] To be sure, there is other evidence that managers believe their "primary" goal should be shareholder profits,[109] but that is perfectly consistent with allowing managers to be influenced by the same social and moral sanctions that influence sole proprietors, who surely are interested primarily in their

105. *See* Clark, *supra* note 1, at 696–703.

106. Id.

107. *See* Choper, Coffee & Gilson, Cases and Materials on Corporations 35 (3rd ed. 1989).

108. *See* Smith, *supra* note 73, at 290–91 (1998); Blair & Stout, *supra* note 80, at 286 n.82.

109. *See* Stephen M. Bainbridge, Corporation Law and Economics 417–18 (2002).

own profits, but not to the exclusion of all social and moral considerations. Indeed, even Dean Clark concedes that in fact corporate managers often assume that they are supposed to temper profit maximization with a concern for other affected interests.[110] Further, whatever managers say they do, empirically corporate managers do not actually profit-maximize, according to many economists, but only profit-"satisfice": that is, they achieve the level of profits necessary to avoid interference with their discretion, but otherwise run the firm to advance other aims.[111]

Thus social and moral factors actually do influence corporate management, making the real question whether corporate law should be structured to minimize the influences of these social and moral processes. My answer is 'no'. An enforceable duty to profit-maximize overrides social or moral sanctions and makes corporations behave in the same way as amoral individuals who ignore the social consequences of their conduct. This would worsen corporate conduct, assuming that our society's social and moral norms do, as a group, improve behavior.

In contrast, managerial discretion to respond to social and moral sanctions will move corporate behavior in the right direction, again assuming our society's social and moral norms correctly identify which direction is right. This remains true even when managers are taking advantage of agency slack to sacrifice profits more than dispersed public shareholders would want. The reason is that the corporate structure weakens the social and moral sanctions applicable to such shareholders, and thus gives them incentives to encourage socially suboptimal corporate conduct.

One might object that many of the social and moral norms currently promoted are misguided or, well, dopey and probably harmfully to the public interest. But I am not saying corporate managers have any duty to respond to every social or moral claim put forth by some group. I am saying they should have some discretion to do so. One thus must distinguish between all the social and moral pressures that are exerted—many of which may be bad—and those to which management yields, which are more likely to be meritorious. Nor am I saying that corporate exposure to social and moral sanctions will always increase the satisfaction of your preferences or mine. I am rather for this point simply assuming that, overall, such exposure would increase the satisfaction of societal preferences, which should be reflected in the full set of social and moral sanctions, even though many individual norms may be questionable.[112]

---

110. *See* Clark, *supra* note 1, at 690–91; *see also* Milton Friedman, *supra* note 1, at 33.

111. *See* Choper, Coffee & Gilson 6th Edition, *supra* note 28, at 29–30 (collecting sources).

112. See the preceding section on "The Social Regulation of Noncorporate Conduct." In the section below on "The Limit That Profits Must Be Sacrificed to Benefit Others" I also address and reject the notion that courts should review whether the particular social or moral norms that influenced managers enjoy widespread support.

Why Excessive Managerial Generosity Is Not a Problem. Assuming that social and moral sanctions on balance are desirable, managerial discretion to respond to them should move corporate behavior in the right direction. However, one might reasonably fear that corporate managers would have incentives to be excessively generous when exercising their agency slack because they bear the full brunt of social or moral sanctions but not the full costs of the sacrifice of corporate profits given that, unlike sole proprietors, they would be sacrificing mainly other people's money. Such incentives for excessive generosity might even push managers so far in that direction that they overshoot the optimal trade-off of profitability and social responsibility. But this is unlikely to be a problem for several reasons.

First, unless the total amount of agency slack were increased, any managerial decision to use their operational discretion to sacrifice corporate profits in the public interest should substitute for profit-sacrificing behavior that would have been more personally beneficial to managers. This seems plausible from the manager side because one would have expected them to fully exploit any agency slack they already have. If they could get away with delivering lower corporate profits, one would expect them to do so by diverting profits to executive compensation, perks, leisure, stock options, or other personally beneficial uses until their failure to deliver higher profits is constrained by other forces. Thus, regardless of any discretion to sacrifice corporate profits in the public interest, one would expect managers to have already gone as far as they could in failing to deliver higher corporate profits. And once they are at that point, then managers cannot simply use such discretion to sacrifice additional corporate profits for public interest causes, but rather have to find some way to offset those lost profits by diverting less to personally beneficial uses. The point is analogous to the familiar point that a monopolist only has a single monopoly profit and cannot just infinitely increase profits by raising prices. Such substitution also seems plausible from the shareholder side because, although shareholders cannot monitor specific operational decisions or determine whether managers are maximizing profits, they can and do monitor the overall level of corporate profitability. Shareholders often won't know whether profits were sacrificed to further personal or public interests or out of sheer laziness or mismanagement, but they do notice declines in profits.

Thus if agency slack is constant, managers who decide to make operational decisions that sacrifice profits to further some public-interest objective will have to make up those profits either by managing the corporation better in other ways (perhaps cutting into their leisure time) or by forgoing other ways of sacrificing corporate profits that benefit managers personally (such as big stock options, fancy offices, corporate jets, or large executive compensation). This means that, unless the amount of agency slack is affected, managers who respond to social and moral sanctions by making profit-sacrificing corporate decisions will be sacrificing "their" profits in the sense of profits that otherwise would have benefited managers or allowed them greater leisure. This would leave managers facing much the same trade-off as a sole proprietor and eliminate any incentive to be excessively generous. Indeed, serious enforcement of a pure profit-maximization standard seems likely to perversely skew manager incentives in a way that makes them more inclined than sole proprietors to advance their personal profits over the public interest. The reason is that under a profit-maximization standard, things like large stock options or executive compensation that help attract, retain, or motivate good managers would be much easier to justify than socially responsible corporate conduct whose connection to profits is more indirect.

Further, if agency slack is constant, any decisions managers made to sacrifice profits in the public interest would leave shareholders financially indifferent, while still advancing the public-interest views reflected in the social and moral norms to which managers are responding. The choice would simply be between paying for that fixed agency slack in the form of overcompensating managers or in the form of corporate compliance with social and moral norms. It is hard to see how the latter choice could generally be undesirable.

Thus the potential problem of excessive generosity cannot arise at all unless there are good reasons to think that managers' operational discretion to sacrifice profits in the public interest would increase total agency slack. And there is little reason to think it would. After all, shareholders cannot avoid giving managers operational discretion and therefore cannot avoid the burden of monitoring it—such operational discretion is a necessary feature of creating an investment vehicle that delegates management to others. The lion's share of cases where this discretion is used to sacrifice corporate profits will reflect latent profit-sacrificing sustainable under the business judgment rule. And most of the rest could be made latent if the law prohibited patent profit sacrificing in the public interest. Such exercises of latent profit-sacrificing authority simply reflect the degree of agency slack that managers enjoy under the business judgment rule; they do not increase it.

The remaining exercises of discretion would involve patent profit sacrificing, where managers do not pretend that the conduct increases corporate profits in some indirect manner. But a rule that allows such patent profit-sacrificing discretion generally does not increase total agency slack, as long as the legal and nonlegal limits on the *amount* of profit sacrificing are the same for patent sacrificing as for latent sacrificing.[113] To the contrary, as I noted above, such patent profit sacrificing tends to reduce agency slack by informing shareholders more accurately about what really is going on.

Managerial discretion to sacrifice profits in the public interest thus seems unlikely to increase total agency slack, and if agency slack is unchanged, then any incentive for excessive generosity is eliminated. The public interest causes benefit, but shareholders do not suffer if any fixed agency slack is exercised in a socially responsible way rather than some personally beneficial way.

113. Those limits are detailed in the section below entitled "Limits on the Discretion to Sacrifice Profits."

Second, even when managers have incentives to be excessively generous, it is far from clear that those incentives would make managers so overresponsive to social and moral sanctions that they have any net incentive to overshoot the optimal trade-off of profitability and social responsibility. The reason is that managerial accountability to shareholders who are underresponsive to social and moral sanctions will create countervailing incentives for excessive stinginess. The net effect may well leave corporate conduct below the optimum—that is, not sacrificing enough profits to further the public interest—despite managerial discretion to sacrifice profits.

For the same reasons, it is unclear whether, on balance, we should expect corporations with managerial discretion to engage in less or more socially responsible behavior than noncorporate businesses. On the one hand, shareholders largely insulated from social or moral pressures should exert pressure through their voting or investment decisions that tend to cause corporate managers to sacrifice profits less often. On the other hand, corporate managers may have some incentives to be more generous to the extent that shareholder accountability is imperfect and the managers are sacrificing profits that do not come out of their own pockets. If corporate businesses are larger and more well known in our modern economy, they also might be more likely objects of serious social sanctions. But whether corporate behavior under current law is more or less socially responsible than noncorporate business behavior is not the question. The question is whether it would improve corporate behavior to change current law by eliminating corporate managers' ability to respond to the social and moral sanctions that help optimize noncorporate behavior.

Third, even if managerial discretion to sacrifice profits does create manager incentives to be excessively generous that are so large that they would cause corporate behavior to overshoot the optimum, that will be undesirable only if managers overshoot that optimum by a margin so great that it leaves their behavior farther away from the optimum trade-off than it would be with a profit-maximization duty. This a conceivable problem, but one that provides an argument against only *unlimited* discretion, not an argument that managers should not have *some* degree of discretion.

Ordinarily, the risk of such excessive managerial generosity is sufficiently constrained not by the law but by product market competition (a firm that takes on excessively high costs cannot survive), labor market discipline (a manager who sacrifices too much in profits will find it harder to get another or better job), and capital markets (the stock and stock options held by managers will be less valuable if they sacrifice profits too much, and this may even prompt a takeover bid).[114] The risk of truly excessive overshooting also likely will be constrained by shareholder voting. Although the shareholder voting constraint certainly is imperfect given shareholders' rational apathy, it should restrain extreme cases of

114. *See* Easterbrook, *Managers' Discretion and Investor Welfare: Theories and Evidence,* 9 DEL. J. CORP. L. 540, 543 (1984).

managerial deviation from shareholder interests. Finally, to the extent that management compensation turns on corporate profits, as it often does, managers will have less incentive to sacrifice corporate profits.

Indeed, proponents of a duty to profit-maximize themselves argue that such market forces will destroy any corporations that do not profit-maximize.[115] But they fail to see that, to the extent they are right about this, it only reduces any need for judicial policing of managerial public-interest activity. In fact, although the product market, capital market, or market for corporate control should constrain excessive managerial generosity, it overstates matters to think that they would produce certain corporate death for any manager who fails to maximize profits. To begin with, managerial profit-sacrificing discretion reflects an agency cost that will be shared by all corporations, like the cost of executive compensation, and thus will not be driven out by market competition.[116] Further, product markets typically are characterized not by perfect competition, but by product differentiation and monopolistic or oligopolistic competition, which gives corporations some discretion to price above cost.[117] Moreover, even where product market competition prevents corporations from raising prices to fund public-interest activities, they can still fund those activities by reducing their rate of return to shareholders.[118] To be sure, this will lower the value of their stock, until the rate of return per share matches other rates of return in the capital market. But this hardly disables the corporation from raising capital. It can just issue more equity at these lower stock prices.[119] Or the corporation can fund reinvestment out of earnings or borrow from lenders to a greater extent.[120] All these strategies will reduce the return to existing shareholders, as well as the long-run ability of the corporation to raise as much capital, but will hardly drive the corporation out of business. And they may not even be noticeable because other corporate managers likely will be exercising the same profit-sacrificing discretion.

The resulting decline in stock price would make it profitable (absent any transaction costs or other obstacles) for a purely profit-maximizing takeover bidder to take control of the corporation and cease its pursuit of nonprofit goals. A perfect market for corporate control thus would make it impossible for corpora-

---

115. *See, e.g.*, Clark, *supra* note 1, at 687–88, 692; Richard A. Posner, Economic Analysis of Law 419–20 (4th ed. 1992).

116. *See* Einer Elhauge, *Defining Better Monopolization Standards*, 56 Stan. L. Rev. 253, 300 (2003).

117. *See id.* at 258, 260; Jean Tirole, the Theory of Industrial Organization 277–303 (1988).

118. *See* Eisenberg, *Corporate Legitimacy, Conduct, and Governance—Two Models of the Corporation*, 17 Creighton L. Rev. 1, 15 (1983) [hereinafter Eisenberg, *Corporate Legitimacy*] (noting that corporations can survive for protracted periods with minimal returns); Coffee, *Shareholders v. Managers*, *supra* note 80, at 20–22, 28 n.76 (collecting sources showing that managers prefer to use internally generated funds and do so for 90 percent of capital expenditures).

119. *See* Lucian Arye Bebchuk, *Federalism and the Corporation: The Desirable Limits on State Competition in Corporate Law,* 105 Harv. L. Rev. 1435, 1466 (1992).

120. *See* Eisenberg, *Corporate Legitimacy, supra* note 118, at 15.

tions to continue pursuing nonprofit goals. But the market for corporate control is anything but perfect. Takeover bidders face enormous obstacles and transaction costs, not only in sheer logistics but in state regulation and corporate defensive tactics.

These various methods of market accountability thus won't entirely stamp out profit sacrificing, but they normally should suffice to prevent excessive amounts of profit-sacrificing. The empirical evidence on corporate donations, which if anything create a greater risk of excessive generosity than do operational decisions, bears this out. The average corporation donates only 1.0 to 1.3 percent of income,[121] which is lower than the individual rate of 1.9 to 2.2 percent,[122] and most of those corporate donations are actually profit-increasing.[123] Thus market forces on average seem clearly able to keep corporate managers from being excessively generous. To be sure, there are special cases where such market forces fail to provide an effective constraint on excessive managerial generosity. But to deal with those cases, the law can impose special limits on profit sacrificing when these market forces are ineffective, as well as a general outside limit on the degree of profit sacrificing. As we shall see, this is just what the law in fact does.

Finally, to the extent that excessive managerial generosity does harm shareholders more than it helps third parties, the cure is worse than the disease for shareholders. Creating an enforceable duty to profit-maximize would, by ending business judgment deference, harm shareholders more than such excessive generosity could. Assuming that business judgment deference has been set to minimize total agency costs, ending it would increase agency costs and thus lower shareholder profits. And unless it were ended, there would be no meaningful reduction in managerial discretion or in incentives to be excessively generous. Further, any profit-maximizing duty would apply not only when managers are exercising agency slack, but also when they aren't. It thus would prevent managers who *are* loyally representing majority shareholder sentiment from profit-sacrificing when that increases total shareholder welfare. This is a particular problem if one wanted to take the minimal step of prohibiting just patent profit sacrificing, for managers are most likely to be open about the profit sacrificing they are doing when they are loyally representing shareholder views.

Those proponents of a profit-maximization duty who acknowledge any such duty is legally unenforceable tend to retreat to the claim that it is nonetheless valuable because it provides a social norm that restrains managers from being lazy or lining their own pockets.[124] But to the extent that this is true, it simply reinforces my points that no legally enforceable duty exists and that social and

---

121. *See* ALI Principles, *supra* note 29, §2.01, Reporter's Note 2; Choper, Coffee & Gilson 6th Edition, *supra* note 28, at 39; Eisenberg, *Corporate Legitimacy, supra* note 118, at 19 n.34.

122. *See* http://nccs.urban.org/chargiving/stgive01_text.pdf.

123. *See* Peter Navarro, *Why Do Corporations Give to Charity?* 61 J. Bus. 65, 90 (1988).

124. *See, e.g.*, Bainbridge, *supra* note 109, at 422–23.

moral norms play an important role in regulating corporate conduct. And even if the supposed profit-maximization duty is really a social norm, that doesn't mean it is the *sole* social norm. It would be just one norm among the larger complex of social and moral norms that regulate conduct. And the analysis in the section on the social regulation of noncorporate conduct still would indicate that pure profit maximization would lead to worse conduct than would decisionmaking that considered both profits and the social and moral consequences of that conduct.

In any event, *pure* profit maximization does not appear to actually be a prevalent social norm. As noted above, investors and corporate managers deny it, economists have argued that managers instead profit-"satisfice," supposed supporting statements say only that the primary objective should be shareholder profits, and even Dean Clark concedes that managers often assume that concerns about other affected interests should temper profit maximization.[125] Nor does a norm of pure profit maximization seem attractive to inculcate. If managerial laziness and self-dealing are the real problem, more targeted social or moral norms against those practices would tackle the problem in a way that is far less overinclusive. A norm of pure profit maximization instead overinclusively demands that managers also maximize corporate profits even when that harms third parties in a way that violates the social and moral norms we traditionally use to optimize behavior.

Why Approval by a Majority of Dispersed Shareholders Should Not Be Required but Approval by a Controlling Shareholder Should Be. The preceding analysis goes beyond showing that the law should not impose an enforceable duty to maximize profits. It also militates against the possible alternative legal strategy of making majority shareholder approval a requirement for public-spirited profit-sacrificing behavior in public corporations. The law justifiably may conclude that investors should not be able, by adopting the corporate form, to render their businesses largely immune from the sort of social and moral pressures that influence noncorporate businesses. Because managers are the only participants in the publicly held corporation who are effectively confronted with social and moral sanctions, they should retain the power to respond to them. Given their insulation and collective-action problems, majority shareholder sentiment will predictably underweigh the social interests implicated. Moreover, even if it were abstractly desirable to require management to obtain a shareholder vote on whether to sacrifice profits, that would not be feasible for the slew of decisions that must be made in the course of ordinary corporate operations about how relentlessly to pursue profits. Thus although the section on how profit-sacrificing discretion can enhance shareholder welfare was correct that majority shareholder approval certainly *suffices* to make managerial profit sacrificing efficient

125. *See* notes 107–111 and accompanying text.

and desirable, it should not be regarded as a *necessary* condition in the case of a public corporation with dispersed shareholders.

The analysis here similarly indicates that encouraging greater disclosure to dispersed shareholders is not an adequate substitute for managerial discretion. Many interesting articles have been written indicating that the law should require managers to disclose whether corporate activities create the sort of harms that raise public-interest concerns commonly held by shareholders.[126] Unless the disclosure were one-sided, it would also presumably require managers to disclose how much profits such activities reap, as well as to disclose any managerial decisions to avoid profitable activities that would have created such harms and how much in profits they sacrificed by doing so. Corporations could also adopt charter provisions requiring such social disclosure. If our only goal were shareholder welfare maximization, such a disclosure strategy would doubtless be beneficial. Indeed, it is interesting that the only shareholder proposals on social issues that tend to come close to obtaining majority shareholder approval are those that seek to require such disclosures. But any disclosure to dispersed shareholders cannot alter the facts that shareholder insulation and collective-action problems will leave shareholders with little incentive to study any disclosed information and quite underresponsive to social and moral sanctions even if they do. Thus no matter how good the disclosure, shareholders in a public corporation likely would favor a suboptimal degree of socially responsible corporate conduct.

On the other hand, where a corporation has a controlling shareholder, that controlling shareholder will be sufficiently identifiable to be exposed to social and moral sanctions, and will not have collective-action problems in acting on them because her decisions can decisively affect what the corporation does. Such a controlling shareholder accordingly should be viewed as the "manager" for purposes of this essay in the sense that she controls corporate operations and is the actor that possesses profit-sacrificing discretion. A lower-level manager should not enjoy discretion to sacrifice his corporation's profits absent some indication of approval by the controlling shareholder of the corporate policy. A sufficient indication of approval generally will exist simply because the controlling shareholder has selected managers who share her corporate philosophy, and requiring an affirmative shareholder vote would be too formal and impracticable given the range of managerial decisions that must be made. But managers should not be able to pursue public-interest objectives secretly or over the known objections of a controlling shareholder.

One might wonder whether the modern prevalence of institutional investors should alter the above conclusions, or at least make the analysis of public corporations more like that of corporations with controlling shareholders. After all, compared with dispersed shareholders, such institutional investors are more

126. *See, e.g.*, Williams, *Corporate Social Transparency, supra* note 86, at 1205–07; Douglas M. Branson, *Progress in the Art of Social Accounting and Other Arguments for Disclosure on Corporate Social Responsibility*, 29 VAND. L. REV. 539, 580 (1976).

likely to be informed about corporate activities, have fewer collective-action problems because they have larger stockholdings, and are less likely to be insulated from social and moral sanctions because they can be identified as a locus for such sanctions. Nonetheless, an enforceable duty to profit-maximize still would be ill advised because it would force managers to ignore the social and moral sanctions that optimize corporate conduct no matter what the institutional investors thought.

Nor does the existence of institutional investors counsel for the alternative of requiring majority shareholder approval. Although more informed and less insulated than individual shareholders, institutional investors remain far less informed and more insulated than corporate managers because they are not directly involved in corporate operations. Moreover, institutional investors have their own collective-action problems because each tends to have a very small percentage of the shares of any particular corporation, and indeed faces legal restrictions against obtaining more than 5 to 10 percent of the stock in any corporation.[127] Thus each institutional investor realizes that its investment decisions are unlikely to affect corporate conduct, and has little incentive to take into account any social or moral impulses it may have. The collective-action problems to exercising their voting rights are high enough that, even when it affects their financial returns, institutional investors spend little effort monitoring corporation-specific policies and rarely make shareholder proposals or solicit proxies.[128]

More important, even if institutional investors did not have their own insulation and collective-action problems, the fact remains that they have to please the individuals who invest in them to obtain their funds. And those individuals who invest with institutional investors are likely to be even *more* insulated from social and moral sanctions than are individual shareholders because those who invest with institutional investors are twice removed from knowledge and responsibility. They may not even know what corporations their investment fund invests in, let alone precisely what all those corporations are doing; and they won't appear on the shareholder lists of any rapacious corporations. For example, individuals who would not dream of investing in tobacco corporations may think nothing of investing in index funds that do.

Even to the extent that individuals who put their money with institutional investors do have social or moral impulses despite their double insulation, their collective action problems give them little incentive to act on such impulses in choosing which institutional investor to invest with. Individuals do not have incentives to choose an institutional investor that conforms to the individual's own social or moral norms when that offers lower returns, for such an individual

127. *See* Mark J. Roe, *A Political Theory of American Corporate Finance,* 91 Colum. L. Rev. 10, 18–23, 26 (1991); Bernard S. Black, *Shareholder Passivity Reexamined,* 89 Mich. L. Rev. 520, 530–31, 542–53, 562–64, 567–68 (1990).

128. *See* Black, *supra* note 127, at 459–60.

decision would have little impact on whether the institutional investor succeeds in advancing that social or moral norm, but would definitely earn the individual lower returns. Consistent with this collective-action problem, even the socially conscious investors who invest in investment funds that commit to social screening have to be induced by commitments that those funds will not actually sacrifice any profits, which necessarily reduces the ability of these funds to have any real impact.[129]

Limits on the Discretion to Sacrifice Profits

The preceding analysis indicates that managers do and should have *some* discretion to sacrifice corporate profits in the public interest. It does not indicate that this discretion is or should be unlimited. To the contrary, it indicates that some limits on that discretion likely will be desirable to prevent any risk of excess managerial generosity.

Limits on the Degree of Discretion

To say that limits are required is not necessarily to say that the limits have to be legal in nature. Normally, no legal limit on public-spirited profit sacrificing is necessary. The discretion to sacrifice profits is instead powerfully limited by managerial profit sharing or stock options, product market competition, the labor market for corporate officials, the need to raise capital, the threat of takeovers, and the prospect of being ousted by shareholder vote. In the lion's share of cases, these market constraints are more than adequate to prevent corporate managers from being excessively generous without any need to employ legal restrictions.

But in some special cases, legal limits do matter. One possibility is that some managers may have idiosyncratic views about the extent of the corporation's social and moral obligations that overwhelm the ordinary disincentives imposed by nonlegal constraints and cause them to make a profit-sacrificing decision that cannot readily be reversed. More typically, legal limits become important when a last-period problem undermines the ordinary effectiveness of the nonlegal constraints on excessive profit sacrificing. Suppose, for example, a manager is retiring. Then none of the market constraints will be meaningful to him because he won't be there to experience them. And if he does something irreversible, such as giving away corporate assets, shareholder voting offers no remedy. The last-period problem posed by retirement normally is not large because it would be rare to have all the managers retire at once, and usually enough managers, including multiple directors, are involved in running the corporation that no single retiring manager can engage in excessive profit sacrificing without the approval of others. Even the chief executive officer, given his pending retirement, will have relatively little ability to get the rest of the board of directors to

129. *See* Knoll, *Ethical Screening, supra* note 87, at 682–84, 692, 710–13, 726.

go along. This is one reason that corporations have multiple directors. Still, it does pose a problem requiring some legal limits.

The last-period problem that typically creates the greatest need for legal limits results when the corporation is up for sale because that can give *all* existing managers a last-period problem at the same time. Because the firm is being sold, the existing managers' decisions about how much to temper profit maximization in the sale no longer will be meaningfully constrained by product or capital markets, nor by the threat of takeover bids or being ousted by shareholder vote. The remaining incentives provided by the labor market, managerial profit sharing, or stock options may be insufficient to constrain excessive profit sacrificing, or may be undermined by buyer (or donee) provision of a new job or special payments to outgoing management. Thus, as we will see, profit-sacrificing discretion is more sharply limited when a corporation is up for sale.

General Limits on Discretion. The law limits profit-sacrificing discretion in various ways. Traditionally, the most common has been to take advantage of the fact that many legal authorities sustained public-spirited activities on the theory that they conceivably maximized long-run profits. Although the business judgment rule meant that in reality these cases gave managers effective discretion to sacrifice profits, the fact that many cases articulated such a test meant that, to be safe, managers had to be able to offer some plausible claim that their conduct could increase long-term profits.

Such business judgment rule review does not actually eliminate profit sacrificing, but it does naturally create a limit on the *degree* of profit sacrificing. In the extreme case, where management operates the corporation in a way that destroys all corporate assets, it clearly would not have any such plausible claim. And in less extreme examples, the more management gives away, the less plausible any long-term profitability claim may be. For example, if management stops clear-cutting even though that cuts corporate profits in half, it will be hard to plausibly claim that the increased goodwill could be large enough to offset this effect. Thus the real constraint imposed by the test requiring a rational relationship to profitability was not that it imposed a duty to profit-maximize, but that it set a limit on the *degree* of profit sacrificing.

However, this traditional approach was not always effective at preserving the necessary discretion, and became much less so once takeover bids became prevalent and monetized how much in profits was actually being sacrificed. Thus corporate law has had to become increasingly explicit that it did mean to authorize some discretion to sacrifice profits, rendering the traditional approach less effective. As it has become explicit about authorizing profit-sacrificing activity, the law has had to use other limits. The ALI does so by saying that managers can devote only a "reasonable" amount of corporate resources to public-interest purposes, and can consider ethical principles only to the extent they are "reasonably regarded as appropriate to the responsible conduct of business."[130] Such a rea-

130. *See* ALI Principles, *supra* note 29, §2.01(b)(2)–(3) & Comments h–i.

sonableness test would constrain management from stopping clear-cutting if it eliminated all profits, and presumably if it reduced profits by 50 percent. But what if stopping clear-cutting reduced profits by 15 percent—would that be reasonable? Alas, conclusory words such as "reasonable" fail to resolve such issues. They serve more as placeholders for standards that either are applied implicitly or one hopes will be provided later. This problem is only exacerbated by the fact that the ALI indicates reasonableness should be determined by considering "all the circumstances in the case."[131] This comes perilously close to a we-know-it-when-we-see-it test.

Somewhat more helpfully, the ALI suggests that the two principal factors to determine reasonableness are: (1) the customary level of profit-sacrificing behavior by similar corporations and (2) the nexus between the public-spirited activity and the corporation's business.[132] But the second factor of nexus does not help at all with operational decisions that sacrifice profits, for such decisions by definition *always* have a close nexus to the corporation's business. The first factor presumably means to capture the notion that shareholders would have expected customary profit sacrificing when they bought their shares and thus will not be harmed by it. Unfortunately, this first factor provides little clarity because there always will be corporations that are above average and below average in their profit-sacrificing levels. If all corporations that exceed the average level are behaving illegally, then half the firms will always be in violation. Presumably, the ALI does not mean to condemn every corporation that donates more than 1.0 to 1.3 percent of corporate income. And if the law stops them from doing so, then the average will keep declining until it reaches zero. The real issue is the degree to which corporate profit sacrificing can exceed this average level, and looking at the average level cannot really answer that question. In any event, customary practice will reflect whatever the legal limits are, and thus cannot tell us what those legal limits should be. We thus have the usual circularity problem that expectations will reflect our legal rule and thus can provide little guidance on what the rule should be.

The underlying problem has been that one cannot articulate a theory that helps determine what degree and nexus of corporate profit sacrificing is reasonable without first establishing a convincing affirmative theory about precisely why corporate management should be able to sacrifice profits at all. With the affirmative theory articulated above, we can begin to make some headway on the issue. The affirmative reason to allow corporate management to temper profit maximization is to subject corporate decisions to the same social and moral processes that apply to sole proprietors when they run businesses. Given that rationale, the appropriate benchmark for determining reasonableness would be the range of plausible behavior for a sole proprietor in the same business position. If the degree of profit sacrificing exceeds what any typical sole proprietor

---

131. *See* ALI Principles, *supra* note 29, §2.01 Comment i.

132. Id.

would do in response to social or moral considerations when he is sacrificing his own profits, then it is unreasonable.

This hardly provides a bright-line test, but it at least provides some guidance about what to look at to determine the extent of profit sacrifice that is reasonable. In particular, because morally devout individuals were historically expected to contribute a tithe of 10 percent of their income to their religious and social communities, one might conclude that managerial decisions to reduce corporate profits by more than 10 percent exceed their reasonable discretion. Consistent with this, cases have held that a helpful guide for determining whether a corporation donation was for a "reasonable" amount is the limit that the tax code sets on the deductibility of corporate donations,[133] which is now 10 percent of corporate income.[134] Likewise, the ALI illustrations indicate that it would be reasonable for a manager to forgo 10 percent of corporate profits by not making a computer sale to a foreign country that would adversely affect national foreign policy,[135] or to forgo 3 to 4 percent of profits by refusing to sell an unprofitable plant to keep workers employed, but it would be unreasonable to do the same act when it sacrifices more than 25 percent of profits indefinitely.[136]

Explicitly recognizing such a discretion to sacrifice up to 10 percent of existing corporate profits seems likely to reduce not just uncertainty, but also the actual extent of discretion that exists under the alternative of a pseudo–profit-maximization standard that allows any action with some conceivable relation to long-term profitability. Because a real profit-maximization standard would undesirably eliminate all discretion to temper the pursuit of profits in the public interest, courts applying a nominal profit-maximization standard tend to accept with credulity any strained claim of a connection to profits, thus leaving management with no clear limits. To the extent courts instead explicitly admit that profit-sacrificing discretion exists and limit it to 10 percent of existing profits, courts can engage in more independent fact-finding and thus be more likely to prevent management from exceeding the 10 percent limit in reality.

Interestingly, this 10 percent standard seems to have an inherent status quo bias. Managers cannot reduce corporate profits by over 10 percent by altering corporate operations. But suppose a timber corporation has always abstained from clear-cutting, and its shareholders all have invested based on the profit stream that policy produces. If clear-cutting would *increase* corporate profits by 20 percent, would managers have an obligation to change corporate policy? No case or ALI illustration appears to have so held or suggested.

---

133. *See* Kahn v. Sullivan, 594 A.2d 48, 61 (Del. S.Ct. 1991).

134. 26 U.S.C.A. §170(b)(2). *Kahn* accordingly sustained a corporate donation of $50 million out of $574 million in corporate income. *See* 594 A.2d at 51, 57, 61.

135. ALI Principles, *supra* note 29, §2.01 Illustrations 21.

136. *See* ALI Principles, *supra* note 29, §2.01, Illustration 19–20. *See also* ALI Principles, *supra* note 29, §2.01, Illustration 6 (making profit-sacrificing loans to needy urban areas is not ethically justified or reasonable when it eats up 24 percent of profits).

Why should this be? If corporate profit-sacrificing discretion is limited by an obligation not to reduce corporate profits by more than 10 percent, why shouldn't it also be limited by an obligation not to forgo decisions that could increase corporate profits by more than 10 percent? For several reasons, it turns out, all of which boil down to the point that the reasons for the former limit do not apply to the latter sort of case. A duty to increase profits by more than 10 percent would be harder to police legally because there would be no historical baseline to turn to: instead courts would have to second-guess managerial judgments about how much profits would be increased by an alternative course of conduct. There also is much less need to police this problem legally because a decision to forgo a change in operations that would *increase* corporate profits, unlike a decision to give away corporate assets, generally will be reversible, and thus much easier to police with standard market forces.

Any duty to increase profits by more than 10 percent would also be less affirmatively justifiable. A 10 percent limit on profit reduction may accurately capture social and moral norms about the maximum tithe-like reduction in individual income. But social and moral norms governing individuals surely in addition prevented them from engaging in rapacious conduct that would have increased their profits by more than 10 percent. And if managers are merely continuing corporate conduct that maintains the existing profit stream, then their decision cannot thwart shareholder expectations or cause a reduction in share price. Finally, any existing pattern of corporate behavior will have reflected the existing set of social and moral sanctions as they have been policed by normal nonlegal constraints. When a manager simply continues that pattern, less reason thus exists to fear either that he has become possessed by idiosyncratic views about the public interest that caused him to alter corporate conduct, or that last-period problems have led to a change of conduct by lifting normal nonlegal constraints.

Thus managerial profit-sacrificing discretion does face a legal limit on decisions that reduce profits by more than 10 percent, but does not face one on decisions that forgo increasing corporate profits by more than 10 percent. One interesting implication of this is that the most important profit-sacrificing behavior will consist not of decisions to reduce profits, but of decisions to forgo the higher profits that could have been made with more rapacious conduct. This is true not only because the legal limits on the latter are looser, but also because we should not see corporations changing their conduct to reduce corporate profits unless there were some change in social and moral sanctions. Thus most exercises of profit-sacrificing discretion will consist mainly of profit-increasing behavior that we *don't* see, but otherwise would have. This necessarily will make most corporate profit sacrificing difficult to observe, especially because a decision to simply continue the sort of corporate activities indicated by unchanging social and moral sanctions might not even be conscious.

The Increased Legal Limits on Discretion When Last-Period Problems Vitiate Nonlegal Constraints. Another strategy the law employs is to alter the legal limits depending on whether management has a last-period problem that makes it less susceptible to nonlegal constraints. When managers have decided to reject a takeover bid and maintain corporate control, managers can consider nonshareholder interests and need not treat shareholder interests as "a controlling factor."[137] Because such managers will continue operating the corporation, they do not face the last-period problem noted above, as any profit sacrificing will continue to be constrained by product, labor, and capital markets, as well as by shareholder voting and their own profit-sharing incentives.

But sometimes takeover bids respond to or result in a management decision to put the corporation up for sale. In such cases, managers do face the last-period problem because they will not continue to operate the corporation. And under Delaware law, the legal standard changes. Where a corporation is up for sale, the important Delaware Supreme Court opinion in *Revlon* concluded:

> The duty of the board had thus *changed* from the preservation of Revlon as a corporate entity to the maximization of the company's value at a sale for the stockholders' benefit. This significantly *altered* the board's responsibilities under the *Unocal* standards. It no longer faced threats to corporate policy and effectiveness, *or* to the stockholders' interests, from a grossly inadequate bid. The whole question of defensive measures became moot. The directors' role *changed* from defenders of the corporate bastion to auctioneers charged with getting the best price for the stockholders at a sale of the company. . . . A board may have regard for various constituencies in discharging its responsibilities, provided there are rationally related benefits accruing to the stockholders. However, such concern for non-stockholder interests is inappropriate when an auction among active bidders is in progress, and the object no longer is to protect or maintain the corporate enterprise but to sell it to the highest bidder.[138]

Likewise, in *Mills Acquisition,* the Delaware Supreme Court stated that managers of a corporation that has been put up for sale, who are assessing various takeover bids, may consider "the impact of both the bid and the potential acquisition on other constituencies, provided that it bears some reasonable relationship to general shareholder interests."[139]

To be sure, some of the *Revlon* language suggests the Delaware Supreme Court thought that normally nonshareholder interests could be considered only when rationally related to shareholder interests, and was pointing out that such a rational relationship could no longer exist when shareholders were being cashed out. But this language apparently just reflects the incomplete waning of the prior

137. Unocal Corp. v. Mesa Petroleum Co., 493 A.2d 946, 955 (Del. 1985). *See also* Ivanhoe Partners v. Newmont Mining, 535 A.2d 1334, 1341–42 (Del. 1987); Paramount Communications v. Time, 571 A.2d 1140, 1153 (Del. 1990).

138. Revlon, Inc. v. MacAndrews & Forbes Holdings, Inc., 506 A.2d 173, 182 (Del. 1986) (emphasis added).

139. Mills Acquisition v. Macmillan, Inc., 559 A.2d 1261, 1282 n.29 (Del. 1989).

doctrine that sometimes grounded profit-sacrificing discretion in such a rational relationship for (as shown above) Delaware case law in fact does not make shareholder interests controlling and thus allows consideration of nonshareholder interests other than just when that happens to maximize shareholder value. When the corporation is being sold, however, management does have last period problems that should make us concerned that they will excessively sacrifice shareholder interests. It thus makes sense to add a special requirement in sale of control cases that any management decision bear a rational relationship to shareholder interests.

Similar language requiring the maximization of shareholder interests does not appear in the Delaware Supreme Court cases about manager decisions to block takeovers or sales of control. Instead, those cases emphasize the discretion of managers to consider nonshareholder interests without limiting such consideration to effects that indirectly benefit shareholders.[140] One of them even emphasized that "absent a limited set of circumstances as defined under *Revlon*, a board of directors, while always required to act in an informed manner, is not under any per se duty to maximize shareholder value in the short term, even in the context of a takeover."[141] Moreover, *Revlon* itself repeatedly emphasized that this duty to profit-maximize was a "change" from the normal duty of managers. So have other Delaware Supreme Court cases applying the *Revlon* duty. They have held that only a sale of corporate control or breakup of the corporation triggers "the directors' obligation ... to seek the best value reasonably available to the stockholders."[142] They have also stated that, outside such a sale of control, managers may base their decisions on the "effect on the various constituencies, particularly the stockholders," but not limited to them, "and any special factors bearing on stockholder and public interests."[143]

Delaware cases also have made clear that, even when a corporation is being sold, it need not simply be sold to the highest bidder. Rather, as *Mills Acquisition* states, management need show only a rational relationship to shareholder interests. Where some of the bids involve a mix of cash and securities, this allows some consideration of nonshareholder interests on the theory that treating them well may in the long run increase the value of the securities shareholders receive. Thus a board can conclude that a bid that looks worse for shareholders at current security prices nonetheless bears a rational relationship to shareholder interests when one considers nonshareholder interests. This was made plain in the *RJR Nabisco* litigation. There an auction was conducted. The winning bid offered a mix of cash and securities, with a face value of $109.00, which the corporation's

140. *See* Unocal Corp. v. Mesa Petroleum Co., 493 A.2d 946, 955 (Del. 1985); Ivanhoe Partners v. Newmont Mining, 535 A.2d 1334, 1341–42 (Del. 1987); Paramount Communications v. Time, 571 A.2d 1140, 1153 (Del. 1990).

141. Paramount Communications v. Time, 571 A.2d 1140, 1150 (Del. 1990).

142. Paramount Communications, Inc. v. QVC Network, Inc., 637 A.2d 34, 48 (Del. 1994); Henry Hu, *Hedging Expectations: "Derivative Reality" and the Law and Finance of the Corporate Objective,* 73 Tex. L. Rev. 985, 1006 (1995).

143. Mills Acquisition v. Macmillan, Inc., 559 A.2d 1261, 1285 n.35 (Del. 1989).

investment banker valued at $108.00 to $108.50.[144] A disappointed rival bidder had offered a similar mix with $3.00 more in cash, for a face value of $112.00, which the corporation's banker valued at $108.50 to $109.00.[145] The Delaware chancery court, in an opinion by Chancellor Allen, sustained the board's decision to accept the first bid, reasoning that the *Revlon* duties applicable in an auction did not bar management from considering nonshareholder interests when the bids are "substantially equivalent."[146] The Delaware Supreme Court dismissed an appeal from this judgment, agreeing that "[n]o legal rights have been established here. The legal issues presented are being addressed by this Court in *Mills Acquisition*."[147]

This conclusion is interesting in two ways. It shows that, even in the auction context, management enjoys substantial discretion because of its power to value bids that include securities. Here that amounted to a discretion of at least 3 percent. Second, the fact is that—even as valued by the corporation itself—the two bids were *not* equal: the accepted bid had a value of $108.00 to $108.50, and the rejected bid a value of $108.50 to $109.00. The corporation's own analysis thus indicated that there was *no* chance the winning bid was worth more than the rejected bid; the best the corporation could say was that the difference in value was between $0 and $1.00. Accordingly, the rejected bid necessarily must have had higher *expected* value to shareholders. The decision effectively holds then that, even in the auction context, management can go beyond considering only those nonshareholder interests that bear a rational relationship to shareholder value. Management apparently can conclude that consideration of nonshareholder interests overrides small differences in shareholder value, amounting to less than 1 percent of expected shareholder value, on the grounds that only "substantial" equivalence is required.

In short, it appears that even under the *Revlon* rules applicable to sales of corporate control, management still enjoys some degree of discretion to sacrifice shareholder profits to further the interests of other constituencies. It need only, if it wants to do so, make sure that the winning bid is structured to include some securities whose value can be claimed to have some rational relationship to

---

144. *See In re RJR Nabisco, Inc. Shareholders Litig.*, 1989 WL 7036, at *1, 14 Del.J.Corp.L. 1132, 1137 [1988–89 Transfer Binder] Fed.Sec.L.Rep. (CCH) ¶94,194 (Del. Ch. 1989) (Allen, C.).

145. 1989 WL 7036, at *2, 9–10, 18; 14 Del.J.Corp.L. at 1138, 1147–48, 1150, 1163. The winning bid was for $81 in cash, $18 in pay-in-kind preferred stock, and $10 in converting debentures; the rejected bid was for $84 cash, $24 in pay-in-kind preferred stock, and $4 in convertible preferred. *Id.*

146. 1989 WL 7036, at *4; 14 Del.J.Corp.L. at 1141 (where "the bids in hand were substantially equivalent in value," the board could accept the bid that "had non-financial aspects that permitted a reasonable person to prefer it"). *Accord* Block, *supra* note 38, at 812 (citing *RJR* for the proposition that "[b]oards conducting an auction ... may consider the interests of non-shareholder constituencies such as employees in choosing between two 'substantially equivalent' offers for control.").

147. 1989 WL 16907 (unpublished opinion reported in table at 556 A.2d 1070).

shareholder value. And it may not even need to do that if the difference in price is less than 1 percent.

However, this degree of discretion still reflects a sharp constriction from the discretion managers normally enjoy to sacrifice corporation profits in the public interest, and the courts seem far more ready to vigorously enforce legal limits in cases involving such a sale of control. This fits well with the theory of this essay, for it is precisely in such auction contexts that management has last-period problems that neutralize normal nonlegal limits on the discretion to engage in profit-sacrificing activities, and thus require tighter legal limits that do not eliminate, but do constrain, that discretion.

The Limit That Profits Must Be Sacrificed to Benefit Others

Another limitation is that profits must be sacrificed in the public interest rather than to further some private interest. By this I decidedly do *not* mean that courts should determine whether the social goal advanced by managers is truly in the public interest. Judges and juries should not be in the business of deciding whether to sustain a management decision, say, to refrain from clear-cutting based on whether the judge or jury agrees that clear-cutting is contrary to the public interest. They have no neutral standards for judging that sort of issue because, by definition, the issue must lie outside the bounds of legal prohibition. Nor do they have any other persuasive basis for second-guessing the views of managers and controlling shareholders. Unlike managers, judges and juries have not been involved in the sort of operational decisions that expose them to social and moral sanctions likely to optimize their behavior. Nor are judges and juries exposed to the market forces that would constrain their decisions. And leaving this issue up to them would make the validity of each corporate decision to sacrifice profits in the public interest turn on the happenstance of which judge and jurors were drawn in after-the-fact litigation, which would be disruptive and fail to provide any guidance for corporate planning.

What I instead mean is that whatever public-interest objective managers cite for the profit sacrifice must involve conferring some general benefits on others, not conferring financial benefits on the managers or their friends and families. Where a corporation does sacrifice profits to financially benefit managers or their intimates, then their decision raises the sort of conflict of interest that vitiates business judgment review.[148] Instead, courts do and should apply the sort of undeferential review utilized under the duty of loyalty, which does actually

148. *See* note 55 and accompanying text. A conflict between a manager's desire to further his public-interest views and the financial interests of shareholders, on the other hand, does not raise a conflict of interest under current law. *Id.* One could imagine calling it a conflict of interest, but that would amount to a general duty to profit-maximize, which would be undesirable for all the reasons discussed in this essay. Indeed, the major affirmative reason for managerial discretion is precisely to allow social and moral sanctions to encourage conduct that conflicts with shareholders' financial interests.

require profit maximization. Because managers in these cases have a conflict of interest likely to bias their decisions, even inexpert judicial assessments about profitability are a likely improvement. Further, because it is limited to cases where managers have such conflicts, such duty of review does not require ubiquitous management by the courts. Thus unlike a general duty to profit-maximize, the duty to profit-maximize in conflict-of-interest cases is one that courts actually can enforce without increasing total agency costs.

One might be tempted to have judges also engage in another form of substantive review—determining not whether the public-interest view held by managers is correct, but whether it is held by enough other persons to reflect some general social or moral norm. The ALI comments appear to suggest that courts should engage in such review when managers decide to sacrifice corporate profits based on ethical principles, with courts sustaining such decisions only when the cited ethical principles are reasonable because they are not "idiosyncratic" or "personal" to the manager, but "have significant support although less-than-universal acceptance."[149] An influential article by David Engel argued for a similar but tougher standard that corporate social responsibility be based only on a clear broad social "consensus."[150]

Such review would help redress the concern that managers might be in a different social milieu than most people, and thus be subjected to social and moral sanctions that cause them to behave in ways that most people would not regard as beneficial. Perhaps, for example, managers run in social circles that cause them to weigh the public-interest advantages of operas and museums much more heavily than the average person would. If so, one might fear that managerial discretion would be exercised suboptimally to spend excessive corporate resources on operas and museums. Allowing managers to exercise profit-sacrificing discretion only when it furthers public-interest views that are widely held by others would help assure that their decisions instead are responsive to social and moral norms that most of us would agree improve behavior.

I doubt, however, that courts really can conduct such a review effectively. To begin with, if they tried to do so, managers simply would camouflage their profit-sacrificing conduct as plausibly profit-enhancing. Courts would not be able to penetrate that camouflage without undermining business judgment rule deference in general.

Even if the profit sacrificing were blatant, courts would have to determine how many others have to hold a social or moral norm to mean that it really reflects a widespread or consensus view of what conduct is beneficial. Such determinations would be hard to disentangle from views about whether the view is normatively correct. Any norm shared by a judge and jury likely will strike them as not idiosyncratic, and any norm they don't hold likely will seem not to enjoy support that really can be called significant. The ALI Reporter, for example, con-

149. ALI Principles, *supra* note 29, §2.01, Comment h.

150. Engel, *supra* note 102, at 4, 27–34.

cludes that a manager could not change a restaurant to a vegetarian menu because vegetarianism does not reflect a generally held ethical principle.[151] But while vegetarianism could not satisfy a consensus standard, it is not at all clear why vegetarianism is not sufficiently widespread to meet the ALI standard of reasonableness. Certainly it seems no more idiosyncratic than a passion for daytime baseball.

Further, if judges and juries were asked whether a view was sufficiently shared by others, they not only would have to assess the numerator (how many held that view) but also would have to make normatively controversial judgments about what the right denominator should be (out of what relevant set of people). This problem has only been increased by the globalization of markets and shareholders. For example, if a Michigan corporation decided to refrain from a profit-maximizing decision to shift jobs to undeveloped nations based on a norm against outsourcing, should courts determine whether that norm is widespread by examining the views of others in Michigan, the United States, or the world generally? Or should courts just consider the views of those actually affected by the decision, or those who have devoted serious thought to the issue, and if so, what constitutes a sufficient effect or serious thought? Any decision about which set of persons should be entitled to determine the relevant norm will necessarily reflect views of the merits.

Even if one could overcome these issues, policing this problem cannot really be done effectively unless courts also reviewed the weight given to the public-interest consideration. After all, divorced from any offsetting considerations, most public-interest propositions would enjoy widespread support and even a consensus. Virtually everyone thinks it better not to clear-cut if there were no cost. Where people mainly differ is on how much weight to attach to the benefits, and people differ on that so extensively that courts cannot simply ascertain a single widespread view, let alone a consensus view. Perhaps recognizing this, the ALI in fact does not try to review the "weight" that managers give to any ethical consideration.[152] But that seems to deprive the review of any significant constraining effect.

Further, when managers devote corporate resources "to public welfare, humanitarian, educational, and philanthropic purposes," the ALI does not purport to review whether those purposes are shared by a sufficient number of other persons.[153] Perhaps the ALI foresaw that trying to do so would embroil courts in normatively controversial judgments about whether funding, say, religion X really advanced the public interest in the views of most people. Courts correctly have declined to get involved in such issues.[154] But because just about any ethi-

151. *See* Eisenberg, *Corporate Legitimacy, supra* note 118, at 11.

152. ALI Principles, *supra* note 29, §2.01, Comment h.

153. ALI Principles, *supra* note 29, §2.01(b)(3) & Comment i.

154. *See* Kahn v. Sullivan, 594 A.2d 48, 61 n.26 (Del. S.Ct. 1991) (rejecting a claim that a corporate donation creating a museum "served no social need" on the grounds that "reasonable minds could differ" about that issue).

cal decision could be reframed as a decision to "devote corporate resources" to some cause (for example, switching to a vegetarian format could be said to be devoting corporate resources to vegetarianism), this means that the ALI standard really does not effectively review whether managers are exercising their profit-sacrificing discretion to further widely held views of the public interest.

Even if administrable, a standard that really required a social consensus on any social and moral norm would likely be undesirable. After all, many laws do not reflect a consensus but rather a majority (and often minority) view that has prevailed over the view of others, yet legal compliance generally is viewed as desirable. The same should be true of social and moral norms that do not reflect a consensus. To assume otherwise is to put a higher burden on social and moral sanctions than on legal sanctions, and thus to bias the conclusion in favor of minimizing the role of the former in favor of the latter. My analysis instead assumes that allowing social and moral norms to influence management decisions is likely to improve corporate conduct because on balance such norms are desirable even without any consensus, or at least they are more desirable than the self-regarding ways in which the inevitable profit-sacrificing discretion of managers would otherwise be exercised.

In short, the only sort of review that courts can and should do about the ends for which profits are diverted is to make sure that profits aren't being diverted to the personal gain of managers or their intimates. As long as managers can show that profits instead are being sacrificed for the benefit of others, that should suffice, assuming that the amount of any profit reduction is within reasonable limits given any last period problem. Courts should not review the merits of the other-regarding purpose in the sense of determining either whether the court agrees with it or whether sufficient others in society do. Limited managerial discretion to sacrifice profits for other-regarding purposes is desirable because, on balance, allowing social and moral norms to influence management decisions is likely to improve corporate conduct, not because judges and juries can pick and choose which social and moral norms are best.

Limits on Which Fiduciary Relations Allow Unauthorized Profit Sacrificing

If corporate managers have discretion to sacrifice corporate profits in the public interest, should that same discretion extend to other fiduciary relations? Should lawyers be able to sacrifice client profits to further public-interest objectives? Should your trustee or personal investment manager be able to sacrifice your money to further some public-interest objective? Should lower-level corporate employees also have discretion to sacrifice corporate funds to further the public interest?

My response to all these questions is that the answer would be yes only if the client, investor, or corporate CEO has approved the profit-sacrificing conduct. Lawyers and investment managers certainly have no duty to engage in rapacious profit-maximizing conduct, even when their clients instruct otherwise. And managerial discretion to sacrifice profits in the public interest could not exist unless upper-level management could authorize such conduct by lower-level managers.

On the other, my answer would be no if no such approval was first obtained. This differs from my answer regarding the profit-sacrificing discretion of managers of public corporations, which I argued above does and should exist even when shareholders have not approved it. The reason for the different answer is that the justifications noted above do not apply to these fiduciary relations. In particular, these cases do not raise the problem of a corporate structure that largely insulates the investor from social or moral sanctions and creates collective-action obstacles to acting on any social or moral impulses. Social or moral sanctions for rapacious profit-maximizing conduct can be visited directly on the client as well as on the lawyer, on the investor rather than on his investment manager, or on the CEO rather than on the lower-level manager. And, as a single actor, the client, investor, or CEO lacks any collective-action problem that would make it difficult for him to respond to such social or moral sanctions. Instead, the situation parallels that between a controlling shareholder and lower-level managers, where, as I noted above, the lower-level manager should not be able to sacrifice profits in the public interest without some indication of approval by the controlling shareholder, who is the best locus of social and moral sanctions. Likewise, in a general partnership where every partner is affected equally by social and moral sanctions, no general partner should be able to sacrifice firm profits without the approval of the other partners. On the other hand, a general partner who runs a limited partnership should be able to sacrifice firm profits in the public interest without the approval of his limited partners because they are likely to be insulated from social and moral sanctions in the same way as shareholders.

Acknowledgments

I am grateful for generous financial support from Harvard Law School and the Olin Foundation, and for helpful comments from Bill Alford, Lucian Bebchuk, Robert Clark, John Coates, David Dana, Jesse Fried, Andrew Guzman, Henry Hu, Christine Jolls, Jerry Kang, Louis Kaplow, Vic Khanna, Reinier Kraakman, Jonathan Macey, Frank Michelman, Martha Minow, Larry Ribstein, Steve Shavell, Lynn Stout, Bill Stuntz, Guhan Subramanian, Al Warren, Elisabeth Warren, and participants in the Harvard Regulatory Policy Program at the John F. Kennedy School of Government, the Harvard Conference on Environmental Protection and the Social Responsibility of Firms, and Harvard Law School's Faculty Workshop, Corporate Lunch Group Talks, Law and Economics Seminar, and Corporate Law Policy Seminar.

Comment on Elhauge

Does Greater Managerial Freedom to Sacrifice Profits Lead to Higher Social Welfare?

John J. Donohue

According to the dominant view of the law and economics literature, corporate managers have a duty to maximize profits within the constraints established by law on behalf of the firm's shareholders.[1] It is recognized, however, that in practice, this duty is not legally enforceable because the business judgment rule affords managers great freedom to govern the firm without significant interference from judges. Any well-counseled manager can successfully insulate virtually any business decision—including the implementation of proenvironmental measures not mandated by law—from attack by invoking the appropriate words designed to suggest the presence of due deliberation with an eye to promoting the long-run interests of the company. Still, the prevailing law and economics conception is that the system will work best if managers are under stress to maximize profits, as this will ensure that they apply themselves diligently to the task of trying to increase the wealth of shareholders. In this view, then, measures that increase product market competition and the threat of hostile takeovers are presumptively beneficial because of the discipline they impose on management. Remove these pressures and managers will work less assiduously while taking greater steps to line their pockets. Those stressing the desirability of increased pressure on managers to maximize profits offer the following empirical evidence in support thereof: the adoption of state antitakeover laws led to clear declines in

1. Although some firms, such as the *New York Times*, are incorporated to pursue objectives other than profit maximization, this is relatively rare. As one of the major law and economics books in the field states, "For most firms the expectation is that the residual risk bearers [i.e, the shareholders] have contracted for a promise to maximize long-run profits of the firm, which in turn maximizes the value of their stock." Frank Easterbrook and Daniel Fischel, *The Economic Structure of Corporate Law* 36 (1991).

stock market values for those firms that did not opt out of these legal regimes, either explicitly or through acts of reincorporation in a state with better corporate law provisions.

But perhaps the traditional view of the obligations of managers of public corporations is wrong. Even when accurate (as opposed to those generated by fraudulent conduct), stock prices may be an inadequate measure of social welfare. In his exceptionally interesting chapter—at once both deft and subtle in presentation, while potentially revolutionary in implication—Einer Elhauge challenges the orthodox conception. He starts out making three broad points that are each at odds, in varying degrees, with the dominant law and economics vision of the obligations of corporate governance of public companies. He argues (positively) that corporate managers do not have a legal obligation to maximize shareholder value; (normatively) that it is desirable for them to have the discretion, influenced by moral and ethical sanctions, to sacrifice firm profits to promote the public interest; and (historically) that the development of an active hostile takeover market imperiled this desirable discretionary capacity but thankfully was corrected by the adoption of state "other constituency" statutes, which clarified that corporate managers did not need to consider the interest of shareholders as dominant or controlling. While the book is focused on the social responsibility of firms concerning environmental protection, Elhauge's message is more general. His comments underscore that from a corporate law perspective, the question of managerial capacity to implement costly and not legally mandated environmental measures is simply a subset of the broader issue of the ability of corporate managers to pursue any corporate action taken in the name of the public interest.[2] Let me address each of Einer's broad points, beginning first with the positive claim that there is no legal obligation to maximize firm profits.

Must Corporate Managers Maximize Firm Profits?

Ask someone strongly committed to environmental concerns if there is a tradeoff between profits and environmental prudence, and one often hears that there is none—it pays to be "green." The implicit message is that corporate actors may not realize that choosing more environmentally friendly policies will be profitable, so furnishing better information should increase the number of "green" choices that will be made. While it should not be surprising that not every privately profitable opportunity to promote the environment will be exploited, the

2. Elhauge also addresses the interesting distinction between socially responsible actions taken in the course of operating the business versus pure charitable contributions. One supposes that, as between operational and charitable acts, pure charitable contributions should be less justifiable in theory, as shareholders would have a greater capacity to supplant the philanthropic judgments of managers if corporate profits were distributed to shareholders rather than contributed to charity. Shareholders would have no comparable ability to choose the socially responsible operational behaviors of the firm, which argues in favor of giving greater freedom to managers in this context.

evidence that corporate executives systematically overlook major cost savings is quite weak. As Elhauge correctly notes, the important hypothetical to consider is not the case in which proenvironmental actions will maximize profits, for then there is no tension between the mandates of a compelling fiduciary obligation to shareholders and a discretionary system directed toward enhancing the extent of pro bono activity. Instead, we must focus on the case in which the trade-off is stark—a manager could act to improve environmental quality in some measurable way, but must sacrifice the profits of shareholders to do so. If it were the case that the managers and directors of public corporations faced a binding *legal* obligation to strictly maximize profits in this situation,[3] then the lesson for this conference and for environmental activists would be twofold: (1) don't ask, or expect, public corporations to sacrifice profits to further environmental or any other public interest considerations (although one could try to point out profitable environmentally friendly measures that management might have overlooked); and (2) target environmental activism toward governmental bodies and toward private businesses that are not run as public corporations.

But what are the *legal* impediments to efforts on the part of public corporations to sacrifice profits for environmental protection beyond levels required by regulation? In addressing this issue, one might begin by asking, "How is a corporation to respond to the regular solicitations it receives from various charitable organizations?" There probably would be some cost to goodwill if the corporation were to reject all such requests, so on pure profit-maximizing grounds, some degree of charitable giving seems both wise and unassailable. Of course, if the law were to take a strong position that such corporate charity is impermissible, then the public condemnation of the ostensible corporate misers would be lessened or eradicated; the corporation could—correctly under that counterfactual assumption—note that it is not free to make contributions.

Certainly, one can find scattered legal precedents for the view that the corporation is not free simply to give firm assets away, and one can use these decisions to fashion an argument that any effort by the managers of a public corporation to deliberately pursue a course of conduct, not required by law, that doesn't promote shareholder wealth falls into this prohibited category. But as one corporate treatise notes, over time "courts have become increasingly liberal and have exercised considerable ingenuity in finding a sufficient corporate benefit to sustain contributions for charitable, humanitarian, and educational purposes, though the benefit might be indirect, long-term, or highly conjectural."[4] The Model Business Corporation Act now expressly recognizes the power of corporations to

---

3. Elhauge's chapter suggests that even absent a binding *legal* obligation, the advent of hostile takeovers in the 1980s meant that firms might have faced a *practical* obligation to maximize profits, in that the failure to do so could lead to a hostile acquisition by a firm that was committed to profit maximization.

4. James D. Cox and Thomas Lee Hazen, Corporations, 2d ed. (2003) at 63–64.

"make donations for the public welfare or for charitable, scientific or educational purposes."[5]

As Elhauge notes, it would be strange, and inefficient, for the law to allow corporations to make charitable donations but to prohibit environmentally beneficial actions that sacrificed similar amounts of corporate profits. But even with this statutory authority, corporations do not have complete freedom to make charitable gifts. A court could bar the making of a charitable contribution if the gift did not bear some relationship to the corporation's present or future well-being. Nonetheless, corporations have frequently satisfied—or circumvented with the aid of the permissive "business judgment rule"—this standard: a Conference Board study found that in 2000, the median corporate charitable contribution was 1 percent of pretax income.[6] It would be interesting to quantify the extent of the operational expenses undertaken in furtherance of some public interest to see the relative magnitudes of the two sets of public interest expenditures—operational versus contributions.

One related point illustrating the development of modern notions of corporate responsibility is that the American Law Institute's (ALI) Principles of Corporate Governance, adopted in 1992 as a form of aspirational restatement of American *corporate* law, rejects the common-law notion that it is permissible for the corporation to *disobey* the law when it is profitable to do so.[7] That is, while corporate managers had greater freedom to break the law to pursue profits under the common law, the ALI opposes this idea, concluding that this one form of managerial discretion should be eliminated. In contrast, Elhauge now champions the notion of greater managerial discretion, but here in the service of going beyond the requirements of law even if profits would be sacrificed in so doing.[8] The same ALI section goes on to permit the firm to "take into account ethical considerations that are reasonably regarded as appropriate to the responsible conduct of business" and asserts that the firm "may devote a reasonable amount of resources to public welfare, humanitarian, educational, and philanthropic purposes." Another treatise on corporate law concludes its discussion of this section with the admonition that "directors and officers are well advised to proceed cautiously if their actions fall into one of the narrow sets of circumstances that are driven by ethical or societal concerns but cannot independently be premised on serving the corporation's long-term economic objectives."[9]

5. Sec. 3.02(13). The same language has been adopted in the single most important body of state corporate law: Delaware corporate law, section 122(9).

6. Amy Kao, Corporate Contributions in 2000, The Conference Board (2001).

7. Section 2.01.

8. To be concrete, consider the case of Guidant, whose subsidiary EndoVascular Technologies recently was prosecuted by the U.S. attorney in San Francisco and agreed to pay $92 million in penalties after pleading guilty to failing to report a faulty surgical device involved in the deaths of 12 patients. Is that behavior something that is more likely to occur if managers feel the pressure of a strong wealth maximization norm or come under greater pressure to pursue overall objectives of corporate responsibility?

9. Cox and Hagan, supra note 4 at 72.

Thus we can offer a pragmatic, qualified affirmative answer to Rob Stavins's question in the Introduction to this volume—is a corporate manager free to pursue the public interest at the expense of profits? As long as a corporate manager doesn't speak too candidly about what she is doing and tailors her remarks in the language of "promoting the long-term interest of the company," she is free to take reasonable, profit-diminishing steps toward more stringent environmental protection efforts. The caveat about not being too candid is a reference to the fact that the one time in American history when a major corporate executive—in fact, the memorable Henry Ford—got on the stand and suggested he was keeping car prices low and wages high simply to promote the public good at the expense of profits, a Michigan court in 1919 instructed him not to run the Ford Motor Company as a public charity and ordered Ford to distribute higher dividends to stockholders. But this is a highly unusual case, not least because virtually no corporate manager today would be as proud and indomitable in resisting the advice of counsel on this point as Ford was, first in a statement to the press and then at trial.[10]

Is It Desirable for Corporate Managers to Have Discretion to Sacrifice Profits in the Public Interest?

In practice, then, corporate managers do have the *ability* to sacrifice profits in the public interest, which is close, but not identical, to saying that they have a legal *right* to sacrifice profits in the public interest. I recall a case in which the government was prosecuting a very sympathetic defendant in a criminal case, and his attorneys asked the judge to instruct the jury about the concept of nullification, in which a defendant is acquitted despite technically satisfying all the elements of the crime. The judge refused to allow the instruction, and the lawyer kept pressing, saying that the jury had the "right" to nullify and should be told of

---

10. In an interview in the *Detroit News*, Ford had stated: "And let me say right here, that I do not believe that we should make such an awful profit on our cars. A reasonable profit is right, but not too much. So it has been my policy to force the price of the car down as fast as production would permit, and give the benefit to users and laborers, with surprisingly enormous benefits to ourselves." In court, Ford was challenged about his remarks.

Counsel: Do you still think that those profits were "awful profits?"
Ford: Well, I guess I do, yes.
Counsel: ... Are you trying to keep them down? What is the Ford Motor Company organized for except profits, will you tell me, Mr. Ford?
Ford: Organized to do as much good as we can, everywhere, for everybody concerned.... And incidentally to make money.
Counsel: Incidentally make money?
Ford: Yes, sir.
Counsel: But your controlling feature ... is to employ a great army of men at high wages, to reduce the selling price of your car ... and give everybody a car that wants one.
Ford: If you give all that, the money will fall into your hands; you can't get out of it.

Ford's lawyers must have cringed at their client's unwillingness to follow the script on his pursuit of long-run profits.

Nevins and F. Hill, *Ford: Expansion and Challenge, 1915–1933*, 97, 99–100 (1957).

that right. The judge replied that the jury does not have the right; they have the power. It would be an invitation for chaos to tell them they have this right, he continued. So the jury was not told, but it did acquit.[11]

While this distinction may be a subtle one, it does apply to our current inquiry. Corporate managers may not have fully a *right* to pursue certain profit-sacrificing measures, but they do have the *power* for exactly the reason that Einer has discussed: the business judgment rule, which is an essential feature of our successful corporate law regime, insulates managers from excessively exacting judicial review of business decisions. As we have seen, as long as the executive is willing to say that the challenged action is being taken as an exercise of business judgment in furtherance of the long-run interests of the company and its shareholders, then it will be largely immune to review.

But whereas some think of managerial ability to sacrifice profits to pursue the public interest as an unfortunate by-product of the desirable business judgment rule, Einer is far more enthusiastic about the beneficial aspects of managerial discretion. Indeed, he finds the primary arguments against such discretion to be rather weak. Einer claims:

> One might reasonably fear that corporate managers would have incentives to be excessively generous when exercising their agency slack because they bear the full brunt of social or moral sanctions but not the full costs of the sacrifice of corporate profits given that, unlike sole proprietors, they would be sacrificing mainly other people's money.[12]

Einer then answers this points as follows:

> Managerial discretion to sacrifice profits in the public interest ... seems unlikely to increase total agency slack, and if agency slack is unchanged, then any incentive for excessive generosity is eliminated. The public interest causes benefit, but shareholders do not suffer if any fixed agency slack is exercised in a socially responsible way rather than some personally beneficial way.[13]

I agree with Einer that if the choice is really between managers diverting shareholder wealth for their personal advantage and diverting shareholder wealth "in a socially responsible way," I would opt for the latter. But that may not be the choice. It may be that anything that gives greater license to the exercise of discretion by corporate managers will simply expand the amount of managerial wealth diversion rather than channel it in a way that leaves shareholders no worse off financially and society possibly better off. I don't mean to categori-

---

11. "Inside the Jury Room," *Frontline* documentary (1987). Had the case been tried to a judge, the defendant clearly would have been convicted, and the judge would have been reversed if he tried to acquit.

12. Nevins and F. Hill, *Ford: Expansion and Challenge, 1915–1933*, 97, 99–100 (1957). Einer Elhauge, p. 55.

13. Elhauge, p. 57.

cally reject Einer's point—only to highlight that it rests on an empirical claim that may or may not be true.

In support of his position, I do offer one minor anecdote. I once had dinner at the house of the manager of a record store. When I commented on his voluminous record collection, he admitted that he would steal a couple of records every time he left the store, but he noted that he was relentless on shoplifters because, in his words, "they cut into my take." Now you might take this as a sign that managers know there is a limit to what they can "steal"—to use a not quite legally precise but at least conceptually illuminating term. Thus, as Einer posits, if you make it easier for them to act in the public interest, it may crowd out more personal wealth diversion rather than encroach further into shareholder wealth.

Einer then makes a very intriguing argument. He states:

> Even when managers have incentives to be excessively generous, it is far from clear that those incentives would make managers so overresponsive to social and moral sanctions that they overshoot the optimal trade-off of profitability and social responsibility. The reason is that managerial accountability to shareholders who are underresponsive to social and moral sanctions will create countervailing incentives for excessive stinginess. The net effect may well leave corporate conduct below the optimum—that is, not sacrificing enough profits to further the public interest—despite managerial discretion to sacrifice profits.... Even if managerial discretion to sacrifice profits does create manager incentives to be excessively generous that are so large that they would cause corporate behavior to overshoot the optimum, that will be undesirable only if managers overshoot that optimum by a margin so great that it leaves their behavior farther away from the optimum trade-off than it would be with a profit-maximization duty.... Ordinarily, the risk of such excessive managerial generosity is sufficiently constrained not by the law but by product market competition (a firm that takes on excessively high costs cannot survive), labor market discipline (a manager who sacrifices too much in profits will find it harder to get another or better job), and capital markets (the stock and stock options held by managers will be less valuable if they sacrifice profits too much, and this may even prompt a takeover bid).[14]

Einer therefore may provide an argument for a greater unleashing of social and moral sanctions on corporate executives in order to correct for the insulation of shareholders from these pressures that the corporate form provides. Of course, that very interesting argument raises two questions: (1) If this argument is true, does that mean we would expect better conduct from a smaller business entity than from a larger corporate enterprise where the social sanctions are arguably weakened, and is there any empirical support for that view? (2) If gearing up a regime to place greater social and moral pressure on corporate executives can influence corporate actions, is there any way to quantify even roughly

14. Elhauge, p. 58.

the costs and benefits of doing so? Will it merely distract corporate executives with dubious popular causes of the day? Might it just provide a cacophony of competing voices so that nothing of substance will be accomplished but costly efforts at window dressing will be undertaken?

It is not hard to concoct a hypothetical to illustrate a situation where all of our moral intuitions would line up solidly behind Einer's thesis that sacrificing profits can promote public welfare. To use an historical example, consider the case of a company manager in 1850 deciding that it was immoral to continue using slave labor in plantation work. Undoubtedly, such a corporate decision would be universally applauded today, although one may wonder whether the social and moral pressures in 1850 would have been more likely to encourage or discourage such a managerial decision. Nonetheless, a publicly traded company making that decision in a world where hostile takeovers were feasible would conceivably be subject to takeover by a firm untroubled by the use of slaves. Such a takeover market would be working as it should in theory—moving assets into their highest-valued use—because shareholder value would be elevated by throwing out the ethical managers. One can see the point behind Einer's claim that enabling greater resistance to takeovers can be a good thing if it protects a socially responsible firm from being pushed into unethical profit maximization.[15] Of course, the simple answer here is that if we get the law right—that is, prohibit slavery—the problem goes away. Note that this suggests that to the extent that we are able to "get the law right," we undermine the definition of corporate social responsibility as embodying a requirement that firms to "go beyond" what the law demands. If we really have the law right, efforts to push the law beyond that will be socially harmful.

But Einer surely would assert, correctly, that we never could get the law so precisely right that individual corporate managers would be unable to enhance social welfare by going beyond the demands of law at times. Law must be constructed in broad categories, and what would be the optimal constraint on average likely will be underconstraining to some entities. Who could complain if unleashing managerial discretion were to make it more likely that managers would take us closer to the social optimum in the environmental or health and safety realm than law alone could get us?

My major concern is not what Einer focuses on in the quote above—that managers may be excessively generous in responding to moral pressures with the money of shareholders that might have different conceptions of the public good—but that anything emboldening managers to exercise more discretion in the expenditure of corporate resources will lead, perhaps perversely, to greater diversion of wealth into the managers' pockets. Indeed, the troubling stories of enormous generosity in making charitable contributions by an array of corporate

15. In theory, one person who doesn't share any charitable impulse could eliminate all corporate donations if you had a costless takeover market (although even in the heyday of hostile takeovers, you needed to have a 50 percent premium before you could justify such a transaction).

miscreants, such as Enron, leads me to offer a testable hypothesis, the answer to which might be quite useful in evaluating Einer's efforts to unleash managerial discretion: "There is a positive relationship between the magnitude of corporate charitable contributions and acts of corporate wrongdoing, including but not limited to managerial wealth diversion."

Was the Adoption of State Antitakeover Law Beneficial?

More than half the states now have enacted laws allowing, and at times even requiring, corporations to consider interests other than those of stockholders. Some even go beyond the traditional list of corporate stakeholders—such as employees, bondholders, creditors, suppliers, and communities—to authorize consideration of the broadest societal interests. In *Revlon Inc. v. MacAndrews & Forbes Holding, Inc.,* 506 A. 2d 173, 182 (1986), the Delaware Supreme Court stated that "a board may have regard for various constituencies in discharging its responsibilities, provided there are rationally related benefits accruing to the stockholders." Other states, such as Indiana, explicitly assert that the interests of shareholders should not be a "dominant or controlling factor" in the decisions of corporate boards.[16]

As one treatise notes:

> The impetus for the states' sudden interest in other-constituency statutes was the hostile takeovers in the 1980s that frequently led to plant closures, layoffs, and declines in the value of the target corporation's outstanding bonds due to the acquisition being financed by issuing debentures to be paid from the target corporation's future earnings .... In this context, the state directive or invitation for directors to look beyond the bottom line loses some of its social responsibility zing, especially under most state statutes, where consideration of other constituencies is couched in precatory language.[17]

Again, one certainly can provide telling hypotheticals about how the environmentally conscious firm will be more at risk of takeover as part of an argument for the desirability of these statutes, and Einer has done a masterful job of making that case. But again, the extremely troubling revelations of serious corporate misconduct and exploding executive compensation in the last 10 or 15 years provide some reasons for concern that something has been going wrong in corporate law of late.[18] If so, what is responsible for this malign set of

16. Indiana Code Ann. Sec. 23-1-35-1(f).

17. Cox and Hazen, n. 1, at 69.

18. A *Financial Times* study examining the 25 largest U.S. public companies to go bankrupt since January 2001 found that their corporate executives "earned" $3.3 billion in salary, bonuses, other cash payments, and share sales between January 1999 and December 2001: "Among the barons of bankruptcy are some familiar names. Ken Lay, former chairman and CEO of Enron, grossed Dollars 247m. Jeff Skilling, former Enron president, grossed Dollars 89m. Even these figures are dwarfed by the Dollars 512m grossed by Gary Winnick of

events?[19] Have the problems been the predictable consequence of an unusual stock market bubble, or are systemic factors enhancing the capacity for corporate misconduct?[20] In particular, we must ask whether state antitakeover laws are partly responsible for, or at least reflective of, the serious problem of managerial excess, or whether they have served the more benign role that Einer suggests of protecting public-spirited managers from sacrificing profits in the service of important social goals.

Consider, along these lines, the case of Armand Hammer, who as CEO of Occidental in 1989 (at the age of 91) persuaded the company's board to authorize roughly $120 million to build the Hammer Art Museum, at a time when Occidental's net profit for that year was $256 million. Shareholders of the company filed two derivative suits in protest. The Delaware Court, though conceding that the stockholders might do well to seek different members of the board, held that their conduct was protected from judicial scrutiny by the business judgment rule. Nell Minow, general counsel of the Institutional Shareholder Services, commented on the decision, stating:

> If the Delaware courts cannot find enough merit in a challenge to a $120 million expenditure for a personal monument that is a twentieth century equivalent to the pyramids to even allow it to go to trial, ... then Delaware does not deserve to be the jurisdiction for these challenges.[21]

Global Crossing, who comes out top of our list." Andrew Hill, "Barons Of Bankruptcy," *Financial Times* (London, England), July 31, 2002, at 1. The study noted that "[g]ains before 1999 and after 2001—often quite large—have not been included."

19. William McDonough, head of the Public Company Accounting Oversight Board, recently noted: "In 1980, the heads of major companies made 40 times more than the average person who worked for them. By 2000, that had increased to 400 times. No economic theory, anywhere, justified that." "Americans Still Furious about Corporate Scandals," *USA Today* at 15A (January 13, 2004).

Ordinarily, one would think that something has changed to stimulate the demand for top executives relative to the supply (which one would assume would be rising fairly briskly with the increase in the educated population and the large growth in the number of MBAs). Moreover, greater product competition from increasingly globalized markets should have acted as a force restraining managerial excesses; however, it also may have bid up prices for better managers.

20. Fed chairman Alan Greenspan recently observed that the distinctive feature of the recent boom was not that people were greedy, but that the "avenues to express greed have grown so enormously." Another interesting question is the extent to which the apparent pathologies in executive compensation are creating potentially politically destabilizing amounts of wealth inequality. As Paul Krugman recently observed, "According to estimates by the economists Thomas Piketty and Emmanuel Saez—confirmed by data from the Congressional Budget Office—between 1973 and 2000 the average real income of the bottom 90 percent of American taxpayers actually fell by 7 percent. Meanwhile, the income of the top 1 percent rose by 148 percent, the income of the top 0.1 percent rose by 343 percent and the income of the top 0.01 percent rose 599 percent. (Those numbers exclude capital gains, so they're not an artifact of the stock market bubble.) The distribution of income in the United States has gone right back to Gilded Age levels of inequality." Paul Krugman, "The Death of Horatio Alger," *The Nation* (Jan. 5, 2004) at p. 16.

21. Minow, "Shareholders, Stakeholders, and Boards of Directors," 21 *Stetson L. Rev*. 197, 213 n. 64 (1991).

Such cases at least raise concerns that efforts to expand the discretion of managers to sacrifice profits to promote "the public interest" can lead to acts whose benefits may not be public and whose consequences may not be so good.

Conclusions

The mainstream law and economics position championed notably by Milton Friedman assumes that corporate managers should try to maximize wealth because the pursuit of this unified objective will enable them to enhance overall social wealth. By passing the profits along to the owners of the company rather than expending them in light of their own preferences, the managers of corporations can facilitate two desirable goals: (1) it will enhance the ability of shareholders to pursue whatever altruistic impulses they have; and (2) it will allow a wealthier society to express its collective choices in furtherance of whatever public goals it wishes to adopt legislatively, rather than leave these choices in the hands of corporate executives who would be spending other people's money without plausibly speaking for the public as a whole.

Elhauge has created a very interesting argument against this traditional view, but given some of the alarming trends in the corporate landscape over the last decade, one has to ask: would unleashing managerial efforts to promote the public good primarily facilitate dishonest or perhaps just self-deluded attempts to further some private agenda, as the Armand Hammer example suggests? Friedman might argue that we already live in the best of all possible worlds if corporate managers believe they have an obligation to maximize profits—a belief that Einer's chapter might subvert—while we avoid all of the problematic litigation that would result if they really did have such a legal obligation. It should not be wholly surprising that if we give them too much discretion, corporate managers would be willing to sacrifice resources at their disposal in disguised efforts to promote their private agendas in the name of the public good.

Comment on Elhauge

On Sacrificing Profits in the Public Interest

Mark J. Roe

In his excellent and complex inquiry, Einer Elhauge argues in his chapter that American corporations should be able to sacrifice profits in the public interest. He observes first that external enforcement is incomplete. Gaps in rules exist and will persist because regulators are not omniscient. If corporate players could fill in these gaps voluntarily, even while sacrificing profits, the public interest could be furthered. Corporate law allows managers to do so (1) indirectly, through the business judgment rule, and (2) directly and explicitly, once we wind through the doctrinal maze. But in recent decades, Einer points out, new transactions, such as hostile takeovers, have pressed American firms to be more profit-oriented, and instability in corporate law doctrine has called managerial discretion into question, as have theoretical law and economics accounts. These accounts and pressures, he argues, need to be rejected.

In this comment, I try to further the analysis along five dimensions. First, the business judgment rule is central. Everything else is a sideshow. The business judgment rule is in fact so large that it effectively allows managers the discretion to sacrifice profits in the public—and any other—interest, without fear that shareholders could successfully sue them. We already seem to be at Einer's doctrinal haven.

But, second, we nevertheless observe a growing emphasis on shareholder profit maximization to the exclusion of other goals. Yet the persisting power of the business judgment rule, and the judiciary's concomitant and continued deference to managerial discretion, suggests to me that the rising emphasis on corporate profitability is probably not due to instability in corporate law doctrine. Corporate law long could stand—and still can stand—the public interest "tithe" that Einer proposes as the norm. The pressure comes from elsewhere, not from corporate law.

Nor is the pressure from hostile takeovers likely to have been the source of new pressure on managers to forgo any preexisting tendency to "tithe." The hostile takeover did in fact strongly reorient the American public firm. But its effect on demeaning a business judgment rule–protected public interest tithe is debatable, or unlikely. The basic financial facts here are twofold: the hostile takeover had its heyday in the late 1980s, but it has largely disappeared since then. More importantly, the typical premium in a hostile takeover—50 percent over the pretakeover price—easily could cover the tithe norm that Einer suggests as appropriate. A tithing manager who did nothing else to induce a takeover normally would not have faced a takeover to prevent that tithe.

But why then is shareholder profit maximization—and its perceived costs—seen to be so important today? Yes, there's more profit-oriented pressure on managers today than there was, say, in the 1950s. But if doctrine still gives managers discretion, if the takeover was likely to be more a managerial excuse than a real reason to pull a prior tendency to tithe, and if ownership structure has been roughly stable, where did the pressures change? The likely culprits are not to be found in instability in corporate law doctrine nor in transactional and financial changes, but in the twin pressures of heightened product market competition and the acceleration of technological change. And the third point here is that not only are product markets and technological change the likely culprits, but that the phenomenon is a worldwide issue, not just an American one.

Fourth, I suggest some considerations relevant to Einer's proposed solution: to give managers more discretion. One issue is that managers, given the business judgment rule, already may be near to having the maximum possible discretion that corporate law could allow. But assuming arguendo that it's possible to increase their discretion under corporate law even further, we still have some issues: if it's the one-two punch from competitive markets and technological pressures that is badgering managers to produce profits, then we might question whether we should shield managers from these pressures. Costs might outweigh benefits; external enforcement of environmental regulation and the public interest, though imperfect, still might be best. And a solution of giving American managers yet more discretion, to insulate them further from market pressures, would give an already favored and powerful section of American society yet more power, discretion, and authority not just to wield their authority inside the firm, but also to make quasipolitical decisions.

Fifth, there's a baseline issue. Perhaps, as Einer posits, the American corporate structure—and not product competition or rapid technological change—whips managers into a profit-making frenzy. Owner–managers, subject to societal pressures, would not be so profit-hungry; but with shareholders distant from the firm, all they care about is profits, the story goes. But it's possible that managers, with their money not at stake to the same extent as owners, are in fact *more* susceptible to social pressures than owner–managers. That at least is what some earlier corporate thinkers, like Thorstein Veblen and John Kenneth Galbraith, thought. In theory, ownership separation could *increase* the impact of nonshareholder norms and values of firms, their management, and their actions far beyond that of the owner-managed firm.

Managers' Discretion

As Einer argues, American corporate law gives managers much discretion to run the company. Even when it's assumed that that discretion should be used for shareholders, courts and corporate law won't question managers' judgment on what's good for shareholders. So managers might give 10 percent of profits to charity, based on the view—or on their assertion—that the burnished public image enhances corporate profitability, by leaving customers with a warm feeling about the company or by better motivating employees. Courts will not second-guess managers' judgment. That discretion under corporate law is sufficient to give managers enough power to tithe the company's profits.

The reader who is unfamiliar with corporate law might be surprised by the corporate law courts' hands-off attitude. Here's a statement of the doctrine from a widely respected business law judge:

> There is a *theoretical* exception to [the business judgment rule] that holds that some decisions may be so "egregious" that liability may follow even in the absence of proof of conflict of interest or improper motivation. *The exception, however, has resulted in no awards of money judgments against corporate officers or directors in [Delaware].* Thus, to allege that a corporation has suffered a loss does not state a claim for relief against that fiduciary *no matter how foolish the investment....*[1]

A tithing manager, even one acting egregiously, just won't yield a cognizable claim to shareholders. The manager is free to act.[2]

What then do courts do vis-à-vis corporate law if they do nothing here?, the reader without a corporate law background might ask. The answer is that courts primarily police interested party transactions, where the managers do not tithe to other interests but, say, "tithe" by putting 10 percent of the company's profits into their own bank accounts.

Moreover, Einer argues, it's not even clear that managers have a formal duty to maximize shareholder profit.[3] Corporate law statutes have fiduciary duties running to the corporation, an entity that comprises more interests than just

---

1. Gagliardi v. Trifoods Int'l, Inc., 683 A.2d 1049, 1052 (Del. Ch. 1996) (Allen, J.) (emphasis added).

2. Doctrinal gaps still might be there. That is, one could imagine a CEO who makes an anonymous charitable contribution in the dead of night, and then quietly reports this contribution to the board, which does nothing. It's hard to see the corporate benefit if it's a secret. Or senior management quietly and secretly protects the environment in a costly way, without even trying to get the public relations benefits. Conceivably a passive board here might be attacked for lacking good faith, for not inquiring, and so on.

3. Indeed, Einer argues that managers clearly *don't* have that duty. Elhauge, at pp. 28–30.

shareholder profits. Because of the strength and breadth of the business judgment rule, I'd suggest, we don't see courts and legislatures explicitly mapping out the limits and beneficiaries of that formal duty. Whether or not corporate law doctrine's permission to managers to act other than for shareholders is wide or strong, the business judgment rule already gives managers all the discretion they'd need to tithe, protect the environment, and so on.

The Takeover Premium

But this discretion has come under pressure in recent decades. It's been under pressure from the quintessential corporate transaction of the 1980s: the hostile takeover. The takeover monetizes managerial discretion, as Einer analyzes. When an offeror is willing to pay 50 percent more than the prevailing stock price, stockholders see how far off managers are from profit maximization. The transaction and its vivid monetary consequences then press all managers to maximize profits, Einer continues, thereby confining the managerial discretion that corporate law doctrine had given them. And a few of the takeover court decisions—especially the early ones—gave some doctrinal strength to a shareholder-centered vision of the American corporation. Although the color from that view faded as antitakeover statutes blanched out the profit-maximizing doctrines from some early decisions, streaks still can be seen from time to time in corporate law decisions, and the view has its academic defenders. Those views, Einer argues, should be rejected. Managerial discretion to sacrifice profits in the public interest should persist, and rules that shield managers from takeovers could be, and should be, justified on such grounds.[4]

That is, shareholders want money. If they don't get it, a hostile takeover will arise (not always, of course, but often enough to make managers jump for shareholders, at least during takeovers' heyday in the 1980s).[5] Therefore, socially responsible managers, who would respect the environment at shareholders' expense, can't. Only if corporate law shields them from takeovers (and the doctrinal encroachments that some academics would like) will they shield the environment from corporate depredations. Thus corporate law's takeover shields are worthwhile. Some antitakeover laws were designed to facilitate sacrificing profits in the public interest.[6] And somewhat restraining takeovers facilitates sacrificing profits in the public interest is therefore potentially wise.[7]

I wonder, though, how strongly the takeover, even in its heyday, really held back managerial social responsibility. That is, Einer's goal is to facilitate a traditional tithe, of about 10 percent of profits. But hostile takeovers typically went forward with 50 percent premiums. A CEO who was costing shareholders 10 per-

4. Elhauge, at pp. 14–15.
5. Elhauge, at esp. p. 13.
6. Elhauge at p. 41.
7. Elhauge at pp. 16, 23–24.

cent of profits so that she could protect the environment would not have faced a hostile takeover just because of that tithe. In other words, I doubt that much of the 50 percent premium in hostile takeovers represented a "monetizing" of managerial profit sacrifice *in the public interest,* but rather a monetizing of managers' *other* profit misses. If pressure to degrade the environment in the name of corporate profits has been rising in recent decades, it probably wasn't pressure from takeovers, or at least not from takeovers alone.[8] It was coming from somewhere else.

Moreover, the profit pressure seems to persist today, although hostile takeovers have largely disappeared, and other takeovers are less frequent as well. So even takeover pressure—never enough to much affect a tithe, I suggest—is largely gone, but the phenomenon of shareholder wealth maximization persists. Something else beyond hostile takeovers must be driving it.

Ownership Structure and the "Baseline"

America's corporate ownership structure impedes social responsibility in another way, Einer argues. Back before ownership separated from control, the American shareholder managed the firm. Back then, before the "fall," owners felt social pressures to act responsibly. Norms vied with profits to determine how the owners acted.[9]

---

8. That is, the typical premium would pay for a lot of clear-cutting, certainly enough to permit the hypothesized tithe (pp. 18–19, 67). The role of takeovers would be relevant to the environment if managers were, say, missing profits by 40 percent for typical reasons of managerial slack, and then their pro-environment policy pushed them into the vulnerable 50 percent region. (More about that below.) The doctrinal risk here is that to get that 10 percent spent for a good environmental policy, we might have to pay another 40 percent in managerial slack. That might or might not be a net public interest gain. If a 10 percent tithe must be paid for by a greater incidence of 50 percent falloffs by managers, two questions are raised. First, is that price too high? Second, does the tithe degrade the public policy result we're after, irrespective of its cost elsewhere?

That is, if firms are less well managed and less societal wealth is created, then fewer resources are available for the ultimate policy goal. So posit that a firm is worth 150 (add zeros if needed) and managers would contribute 6.7 percent to projects that are truly socially beneficial and not in shareholders' interests. But to get to a full tithe, we'd need takeover defenses. Those defenses though would leave managers free to let firm value drift downward to, say, 100. They tithe, but at 10 percent of 100. If full tithing has to be done with takeover defenses, then on these numbers not only is the society wealthier overall without the tithe-plus-antitakeover package, but also the policy goal sought with the tithe will *itself* be just as well funded: $.1 \times 100 < .067 \times 150$. The tithe package is 10 percent of, say, a firm worth 100. But without takeover defenses, the firm is worth 150 and at 6.7 percent, a little more than 10 ends up being spent on the desired policy.

A second question of circularity arises. Is the 50 percent premium the result of antitakeover law? In the absence of antitakeover law, would the premium equal zero? Probably not, because the high premium preexisted the strong antitakeover laws of the late 1980s. Subsequent to these laws, the frequency of takeovers declined as a result of either the laws or other economic reasons.

9. As Einer says, managers should "subject corporate decisions to the *same social and moral processes that apply to sole proprietors* when they run businesses." Elhauge, at p. 67.

But today's ownership structure whips managers into a profit-making frenzy. Distant shareholders don't understand how their firms affect the environment. Financial managers at institutional investors are judged on their ability to get money into shareholders' pockets. Corporate managers respond to investors' goals by single-mindedly pursuing profits. That's the story in the paper, and it's a common one.

Certainly that inequality—owner–managers being more susceptible than distant shareholders and professional managers to societal norms—is plausible. Separation might weaken the impact of social norms on the owner–manager nexus, as is often assumed.

Or separation might *augment* the impact of social norms on corporate decisionmaking. Owner–managers, after all, were once called robber barons, presumably because of their single-minded pursuit of profits, often in ways that degraded the public interest. As a matter of intuition, it's plausible that the owner–manager, with his own money on the line, would pursue profit maximization *more* single-mindedly than the professional manager. Indeed, earlier analysts of the corporation, such as Thorstein Veblen and John Kenneth Galbraith, viewed separation as an *opportunity* for social norms to affect corporate decisionmaking *more* intensely than when the owner–manager—the robber baron—dominated the American corporation.[10] Professional managers would not run the company solely with shareholder profits in mind. They would pursue a social interest. And social interests would affect the professional manager, who, when directed by legislation or regulation to do this or that, would comply more readily than the owner–manager. In this perhaps older view, separation was a *solution*, not a problem. If that old view is right, then we haven't "fallen" from an old state of grace.[11] And today the most active institutional investors (public pension funds, AFL-CIO) are the least likely to be solely profit-oriented and the most likely to look to wider social values.

Possibly there's no one right view here. Sometimes the owner–manager would feel the weight of social norms and, already rich, give up some shareholder profits. Sometimes the professional manager, having displaced the robber baron, would be more sensitive than the robber baron to the firm's social impact. The professional manager might get social accolades from friends and family, yet it would be the distant shareholders who paid.[12] The situation might vary from industry to industry, from owner to owner, from one manager to another, or

10. Thorstein Veblen, The Engineers and the Price System 70–72 (1936); John Kenneth Galbraith, The New Industrial State 81–82 (4th ed. 1985, orig. 1967).

11. Moreover, with ownership separated from control, and with managers able to sacrifice profits up to 50 percent before a hostile takeover kicked in, then maybe Veblen and Galbraith wouldn't be that far off from describing today's reality.

12. Arguably a social benefit to a director—in the firm directing resources to the director's preferred charity in return for an understanding that the director wouldn't challenge the managers' corporate actions—is the type of benefit that could be actionable in court, but

from time to time. But as a matter of logic and observation, it's not clear and certain that it's the owner–manager who is always the ideal type that we'd want social policy to replicate.

All around the World

Let's stick for the time being with the view that America's ownership structure whips managers into a profit-making frenzy: shareholders are too distant from the firm to see the damage done; they get information only on financial performance and managers are rewarded only for financial performance. So shareholders press managers to return dollars to them, not to maximize social utility.

But strong profit pressure on the firm is not just an American phenomenon. In other nations where firms' ownership structure more resembles that of the hypothesized socially responsible owner–manager, the pressure to maximize shareholder profit is there, and has been rising. Consider just the titles of the following articles:

- Bradley and Sundaram, The Emergence of Shareholder Value in the ***German*** Corporation (SSRN, Oct. 2003).
- Rose and Mejer, The ***Danish*** Corporate Governance System: From Stakeholder Orientation towards Shareholder Value (SSRN, 2003).
- Joerg, Loderer, and Rothe, Shareholder Value Maximization: What [***Swiss***] Managers Say and What They Do (SSRN, 2002).
- Bilefsky, [***European***] Shareholders Are Demanding Greater Accountability from Business Leaders, Wall Street Journal, Nov. 24, 2003, at A10.

And these nations' corporate law doctrines and structures are more susceptible to the hypothesized socially responsible owner–manager than are American corporate law and American corporate structures. Some nations explicitly demean shareholder profits in their corporate law doctrine, looking for firms to pursue the national, or the social, interest. Some nations bring other interests into the boardroom, as Germany does via codetermination, which bars both managers and shareholders from dominating the German corporate boardroom, because German corporate law requires that employees get half of the board's seats. And owner–managers and family-owned firms are much more common abroad.

Yet despite the absence of American corporate doctrine, American takeovers, and American ownership structures, the pressure to make profits is said to be increasing all around the world. What accounts for that pressure?

typically hasn't been the focus of lawsuits. Such relationships are real, though, and courts are now getting around to realizing their importance in co-opting outside directors. See In re Oracle Corp. Derivative Litigation, 824 A.2d 917 (Del. Ch. 2003) (Strine opinion on Stanford connections that potentially allowed insider to co-opt outside directors).

Perhaps the answer lies not in the corporation, its structure, or its governing rules, but in its environment. Product market competition has intensified in the past five decades. Technological change is more rapid than it once was. These two phenomena—the core phenomena of globalization—may account for much of what Einer observes. When product market competition is severe, managers cannot readily forgo much in the way of corporate profits without facing a competitive penalty. And if technological change is swift, managers cannot easily spend too much time thinking about subtle environmental consequences or they could miss the next technological leap and then be washed out from the market. They can't lose focus or they might lose out. These product and technological pressures squeeze all managers—even managers with corporate law discretion, even managers free from hostile takeover pressures, even owner–managers who are savvy and sensitive to social consequences.

Back, say, in the 1950s, when oligopolistic competition was more common, when technological change was slower, and when American industrial managers didn't face much competition from Japan, Europe, or China, American managers could more readily look beyond profit maximization.[13] They had the doctrinal freedom to do so under corporate law. They had slack that they could use for their own benefit, and sometimes for the public benefit. When competition—international and domestic—intensified in the ensuing decades, and when technological change accelerated, American managers lost much of the discretion that they once had.[14]

Thus we have four competing explanations for the phenomenon of intensified managerial attention to shareholder profits: a corporate law explanation, arising from recent instability in corporate law's traditional discretion for managers; a corporate transaction explanation, driven by takeovers; a corporate ownership explanation, coming from distant shareholders whipping managers into a profit-making frenzy; and a product market cum technology explanation. All could be in play, although I'd put my money on the last combination as likely to be the strongest of the four. And if it is the strongest, then reforms to further insulate managers from shareholders—even if sound otherwise[15]—would advance their public-regarding goal only weakly, because strong product and technological pressures would persist.

13. Edward Mason, ed., The Corporation in Modern Society (1959).

14. Mark J. Roe, *From Antitrust to Corporate Governance*, in The American Corporation Today (Carl Kaysen, ed. 1996). Even when technology yields a monopoly, the pace of change has been so rapid recently that managers know that the monopoly is likely to prove transient unless they improve and compete against the next new thing.

15. See the next section.

More Discretion?

If corporate inattention to environmental responsibility is the problem, is more discretion to managers—rather than, say, more enforcement resources or more penalties on offending managers—the solution? Or, similarly, is buttressing their already wide discretion with a further powerful justification wise? More discretion, although plausible as the solution, or a solution, has costs—in corporate governance generally—and might not be as effective as it at first would seem likely to be.

The best case for more discretion to sacrifice profits would arise if managers have a fixed packet of authority, one that doctrinal change in the public interest wouldn't expand. If the size of managerial discretion is already fixed and not expanded, then the profit sacrifice would, in theory, have managers act more in the public interest than they have been. And indeed, the wide discretion that the business judgment rule affords suggests that we can't increase discretion further. But this possibility has several problems. First, if managers' packet of divertible resources remained constant, then presumably we'd be hoping for managers to substitute the public interest for their own benefits, or substituting a new way to promote the public interest for a prior one—one that had been justified via the business judgment rule but now could "go direct." Substitution for their own interest or profit-justified acts is surely possible, but we might question how likely, deep, and important that kind of substitution would be.

Second, consider the alternate possibility that there's room to expand managerial authority here, that the business judgment rule doesn't already give American managers a free hand. If so, keep in mind that enhanced managerial discretion would give managers authority to bring about *their* vision of the public interest; they need not implement the experts' cost–benefit analysis or the public's consensus. For example, managers might think clear-cutting (Einer's bête noire) is good—so that, I suppose, grasslands can dominate and cattle can run free. James Watt, after all, pursued a vision of the public interest in the environment. It just wasn't the typical environmentalist's vision.

There's some history here that would trouble the typical, committed environmentalist: in an early run at such problems, managers at Dow Chemical were attacked by public interest groups for the company's manufacture of napalm in the 1960s. Dow managers defended their pro-napalm policy, not on the grounds that making it was what Dow had to do to maximize shareholder profits, but saying, to the contrary, that making napalm—never a profitable product line for them—was the socially right thing to do.[16] Managers' vision of the public inter-

16. Medical Comm. for Human Rights v. SEC, 432 F.2d 659 (D.C. Cir. 1970) ("Dow [management] ... proclaim[ed] that the decision to continue manufacturing and marketing napalm was made not *because* of business considerations, but *in spite* of them; that management ... decided to pursue a course of activity which generated little profit for the shareholders and actively impaired the company's public relations ... because management considered this action morally and politically desirable.") (Emphasis in original).

est may not match the experts' vision, the activists' vision, or the public's vision of the public interest.[17]

Expanding managerial power, again assuming that one could expand on the business judgment rule, would raise other issues. Managers might exercise their enhanced, or better-justified, discretion not to benefit the public, but to indirectly benefit themselves by, say, co-opting outside directors, thereby making the outsiders more likely to be in the insiders' pocket on other corporate issues.[18] More discretion might yield more power to insider managers to co-opt outside directors, a problem that is not small today and is part of the core of the recent Enron and other scandals.

Moreover, we should think twice about whether we should expand the discretion, if we could—or further justify it, if we can't—of an already powerful group in American society, giving managers quasilegislative authority. And, lastly, if more discretion were possible, then the issue of the congruence of the managerial vision of the public good with others' visions of the public good—the visions of the cost–benefit analyst, the regulator, the activist, and the public—would be even more strongly in play.

Conclusions

Einer Elhauge's chapter is an excellent inquiry into our understanding of shareholder profit maximization and the consequential corporate law issues. This is not an easy topic, and he is mindful of the difficulties and complexities, which are many and may be why most academics shy away from this kind of inquiry.

I offer a few extensions and reservations. First, the business judgment rule is probably central, already giving managers ample discretion to tithe profits in the public interest. Yes, takeover and ownership structure might affect this discretion, but that conclusion is not inevitable. As hostile takeovers typically pro-

17. Einer recognizes this problem. A limit, he points out, is that shareholders can vote managers out. (48). Three features make that limit not a very tight one. First, we know how hard it is to vote managers out; that's why there was a 50 percent premium in the hostile takeover. Second, the goal in Einer's chapter is to make sure managers have that discretion, to free themselves from profit-hungry shareholders, who, Einer points out, are structurally free from social and moral pressures (p. 4). To the extent that managers are justifiably free from shareholders, they have discretion to act against environmental interests (as environmentalists see those interests), as well as for them. Third, shareholders' goals might also differ sharply from those environmentalists hold.

18. Sapping director independence via subtle compromises is not a minor, theoretical issue. Although the Enron board, for example, was for the most part formally independent of management in terms of straight financial conflicts, "charitable donations and political contributions created relationships between management and the Board that may have weakened the independence and objectivity of certain Enron directors." See Charles M. Elson & Christopher J. Gyves, *The Enron Failure and Corporate Governance Reform*, 38 Wake Forest L. Rev. 855, 872-73 (2003); Stuart L. Gillan & John D. Martin, Financial Engineering Corporate Governance and the Collapse of Enron (U. Del. Center for Corp. Governance, WP2002-001, 2002) (board independence weakened by Enron's "donations to groups with which directors were affiliated").

ceeded at a 50 percent premium, they would not seem likely to bar a tithe. And the hostile takeovers have largely disappeared. The ownership structure argument—that distant shareholders care only for profits, and managers fall into line with shareholders' goals—is plausible but not necessarily certain. Earlier thinking here was to the contrary: that owner–managers—the robber barons—*demeaned* the social interest, and that separation *promoted* it, by allowing societal norms to affect professional managers, by slowing down the profit-making machinery, and by creating a corporate structure on which government policymakers could act more effectively.

Moreover, the phenomenon of an increasing shareholder orientation in recent decades is a worldwide one. But other nations are less affected by American-style ownership separation, hostile takeovers, and the other American institutions involved. The primary culprits in badgering managers to produce profits may turn out not to be corporate law instability, hostile takeovers, or ownership structure, but the worldwide—the global—intensification of product market competition and the acceleration of technological change. If that's so, then efforts to reform corporate law may have limited impact, because the powerful pressures toward profits are coming from elsewhere.

And if social engineering by tinkering with corporate doctrine could have an impact, would it be a good idea? In any ledger of pluses and minuses, one would have to weigh in some of these minuses: we'd be enhancing the authority of an already powerful sector of American society, formally giving it quasilegislative power. We might be further justifying managerial insulation, with costs in corporate governance, without being sure we'd get the big payoff in better social policy. For example, if we'd get the social tithe by shielding managers from the remaining small chance of a hostile takeover at a 50 percent premium, we might have to pay too much in corporate profits—much of which ends up in the U.S. Treasury for public purposes and the rest of which, even if private, is not always entirely socially disreputable—to get that tithe. Maybe it's worth it. Maybe it's not.

Summary of Discussion on Corporate Social Responsibility and Law

This discussion focused on three questions. (1) *The positive question*: To what extent does the law formally permit corporate managers to sacrifice profits for the sake of social responsibility? (2) *The normative question*: To what extent should we permit or encourage corporate managers to sacrifice profits for the sake of social responsibility? (3) *The relevance question*: In light of other factors constraining corporate managers, does it matter whether we think they can and should sacrifice profits for the sake of social responsibility?

What the Law Permits

On the positive law question, Robert Stavins asked whether it was truly noncontroversial that corporate managers have no enforceable legal duty to profit-maximize. Einer Elhauge replied in part that any lawsuit brought against a manager on this basis is almost certain to fail. He was supported in this view by Mark Roe and John Donohue, both of whom agreed that lawsuits of this type are doomed to fail. Roe noted that the business judgment rule is extremely broad. Direct fiduciary suits will not succeed; rather, the "duty" to shareholders arises more from legal structure than from lawsuits. Donohue added that he thought there were limits on a manager's freedom to sacrifice profits explicitly. He argued that if a manager of a public company announced a plan to sacrifice substantial profits solely for public-interest reasons, courts might step in to impose a fiduciary duty. But because managers face other constraints against explicitly sacrificing profits, and because they always assert that their acts that appear to sacrifice profits really will maximize profits in the long run through enhanced goodwill, there are no cases on point.

Bruce Hay argued that the positive law issue really decomposes into two questions: what the manager's legal duty is and, whatever it is, how rigorously the courts will enforce it. He pointed out that even if the manager's duty is to maximize profits, the duty might not be—and in practice is not—strenuously enforced by the courts, because enforcement requires judges to evaluate business decisions, which they are ill suited to do. Hay stated that courts are highly deferential to managers on issues of business policy, for the same reason that they are highly deferential to agency administrators on issues of regulatory policy that have been delegated to the agency. For example, if Congress instructs an agency to issue economically optimal regulations, a court will be highly reluctant to substitute its views for the agency's on whether a given regulation is optimal, because the court lacks the necessary expertise. So, too, Hay contended, no court will substitute its judgment for that of a company's managers on what is privately optimal for the firm's shareholders. That, he noted, is the essence of the business judgment rule: even if a manager's obligation, in theory, is to act in the best interests of shareholders, courts will not second-guess managers' business decisions. The lawyers at the conference generally agreed with this assessment.

Several of the business and economics scholars indicated that they were surprised to learn that managers lacked a legally enforceable duty to maximize profits. Two of the legally trained participants suggested, however, that if business schools are teaching that managers should be working for shareholders, this result still could be justified. Cary Coglianese stated that there might be value in a myth of duties owed to shareholders in order to keep managers focused on the bottom line and to keep them from diverting corporate assets for their own benefit. Roe said that although business schools ought not to maintain there is a duty, business schools still might want to use a principle of profit maximization as an instructive shorthand for future business leaders, who will face intense product market competition, rapid technological change, and shareholder scrutiny. Shareholders elect directors. Even if they cannot sue them for sacrificing profits, shareholders can in principle replace them. Although elections are not tightly under shareholders' control, it is fair to say that shareholders will not be pleased with directors and managers who sacrifice profits.

What the Law Should Permit

Regarding the normative question, Stavins and Daniel Esty asked why one should expect corporate managers to make socially beneficial decisions about profit-sacrificing activities. Esty pointed out that when it comes to CSR, corporate managers have extremely broad discretion, ranging from environmental actions to opera donations. Stavins put forth the hypothesis that society is better off if managers focus on maximizing profits, subject to constraints imposed from outside authorities, such as environmental regulators. He therefore suggested that it might be socially inefficient for managers to deliberately sacrifice profits when not required to do so by such external constraints. Why, he asked, should we expect corporate managers to choose profit-sacrificing projects that maximize social net benefits? When a firm imposes costs on itself for CSR, those costs result in higher prices that decrease net benefits in its product market. Those costs might be greater than the benefits that come from correcting the externality. Is there any theoretical or empirical argument to support the claim that managers will make these trade-offs efficiently? Stavins further noted that society already regulates some environmental pollutants above the efficient level of control, so CSR with respect to such pollutants almost certainly will be socially wasteful.

Elhauge responded first by arguing that managers are often in a better position than regulators to make efficient trade-offs about case-specific issues, because as direct participants, they have better information. Informational limitations, he maintained, necessarily will lead to imperfect legal rules, leaving room for discretionary decisions that improve on the behavior required by legal rules. For this reason, societies have always supplemented legal and economic sanctions with moral and social sanctions that take advantage of information possessed by those engaged in conduct. While agreeing with Stavins that regulations often do have excessively stringent applications, Elhauge pointed out that managers already must obey the law and are not likely to feel much moral or social motivation to exacerbate what, given their better information, they know is an overinclusive legal application. Instead, he argued, moral and social sanctions are more likely to alter discretionary managerial behavior where the regulations are too lax and managers know that profit-maximizing behavior would undesirably exploit this legal underinclusion. This effect of moral and social sanctions on discretionary managerial behavior will be desirable, according to Elhauge, unless it causes managers to overshoot the optimum by a wider margin than the shortfall that would result if they did nothing but maximize profits in compliance with inevitably underinclusive law. The risk of overshooting, he held, is likely to be constrained by many factors, including product markets, capital markets, labor markets, takeover threats, shareholder voting, and managerial profit sharing and stock options.

Ultimately, Elhauge argued, the question is how well structured our system of moral and social sanctions is. If social and moral sanctions are well designed, then allowing managers to respond to them by giving them the discretion to sacrifice profits in the public interest will have desirable results. If moral and social sanctions are poorly designed, then allowing managers to respond to them could instead worsen behavior. Elhauge concluded that it would be startling if social and moral norms were so inaccurate that managerial behavior would be improved if managers had an enforceable legal duty to ignore the moral and social ramifications of any effects their conduct might have on third parties. He observed that such a conclusion also would suggest that the behavior of everyone would be improved if all had a legal duty to ignore the effects of their conduct on third parties unless it ultimately redounded to their own financial gain, which would be difficult to square with much literature on moral philosophy.

Donohue supported Elhauge's point by emphasizing that government regulation ordinarily takes the shape of uniform standards imposed on firms that differ in their compliance costs. The upshot, he noted, is that firms with low compliance costs frequently are underregulated. To the extent that CSR encourages managers to correct the resulting inefficiencies, Donohue remarked, it is socially beneficial.

Paul Kleindorfer argued that in some cases, desirable environmental outcomes can be brought about more cost-effectively through the social norms leading to CSR than through traditional regulation. When firms change their behavior to accommodate social norms—catalyzed, perhaps, by reporting requirements—they eliminate the transaction costs entailed in monitoring, enforcing, and litigating traditional regulation. He drew an analogy to the control of litter: the government does not have the resources to monitor and enforce laws against littering, but it can control the problem more cost-effectively by promoting social norms against the practice.

The Relevance of Law

The final question taken up in discussion was how much it mattered whether the law permits CSR, given the financial pressures on managers to maximize profits. On this issue, a number of conference participants called attention to recent economic and legal developments that have expanded the possibilities for CSR in some respects, while constricting them in other respects. Paul Portney contended that managers may have more latitude than they did in previous periods, because a greater percentage of corporate assets have become intangible. In the 1980s, about 60 percent of assets held by Fortune 1000 companies were tangible. Today that figure is less than 20 percent. This change has complicated the process of measuring profits, making it harder for the market to "punish" managers who deviate from the profit-maximization goal.

Donohue claimed that other developments have exerted a contrary pressure, making CSR less likely to occur. He pointed to the rise of stock options as a form of managerial compensation in the past 15 years, which has increased managers' incentives to concentrate on shareholder value. Donohue also echoed Roe's argument that heightened product market competition hampers managers' ability to sacrifice profits for the sake of other goals.

David Vogel argued that students of CSR should pay more attention to factors constraining managers' ability to behave in environmentally responsible ways. His argument was directed at the growing literature that highlights the pressures on managers to engage in CSR—pressures such as social norms, investment funds that pursue social objectives, and consumer behavior that rewards environmentally conscious firms. He argued that this literature paints a distorted picture, because it ignores the fact that increased global competition has pushed managers to behave in a less socially responsible fashion.

Kleindorfer countered by pointing to empirical work done by himself, Dennis Aigner, Kip Viscusi, and others finding that firms do in fact respond to social norms and community pressure. In some cases, corporations enhance their profitability in doing so; in other cases, they do not. Over time, he argued, one would expect best practices to permeate and the least beneficial practices to disappear.

PART II

The Economic Perspective

Paul R. Portney

Comments

Dennis J. Aigner

Daniel C. Esty

Corporate Social Responsibility

An Economic and Public Policy Perspective

Paul R. Portney

The term "sustainable development" began popping up prominently in public policy, and especially environmental debates, beginning with the publication in 1987 of the report "Our Common Future," by the World Commission on Environment and Development.[1] Similarly, over the last 10 years or so—and at least in part as an offshoot of the ongoing debate about sustainable development—we have begun to hear more and more about what has come to be known as corporate social responsibility (CSR).[2] As pointed out by the World Business Council for Sustainable Development, no well agreed-upon definition of CSR exists apart from a commitment by corporations to contribute to sustainable development, itself a remarkably ambiguous term.[3] Nevertheless, conferences on the subject abound, and each passing week seems to bring with it strong statements by environmental and social activists that companies must do more to adhere to the principles of CSR, as well as responses by companies touting their commitment to the same.

Public interest groups and business advocates can speak for themselves on the merits and shortcomings of CSR. But how should those interested in enlightened public policymaking view it? Do firms engage in a variety of activities that are consistent with CSR? Is there evidence that it is in their interests to do so? Does it make sense to encourage firms in the private sector to embrace CSR? What evidence is there that doing so would redound to society's benefit? Or might it be

1. World Commission on Environment and Development (1987).

2. The debate about the proper role of the corporation in modern society goes back much farther than 10 years, of course. However, casting this debate as one about CSR—and particularly a corporation's environmental record—is a more recent phenomenon.

3. World Business Council for Sustainable Development (2000), p. 10.

the case that doing so could actually make things worse, as strange as that may seem? How would we know if a firm has "overdone" CSR, if that is possible? It is the purpose of this chapter to answer these questions.

In the next section, I provide a short definition of CSR, without which it would be virtually impossible to proceed. I turn then to a very brief discussion of the traditional rationales, on grounds of economic efficiency, for government intervention in an economy where the provision of most goods and services is left to individuals and private firms. This is necessary to any discussion of the welfare effects of CSR, one of the principal purposes of this chapter. Following that, I consider the possible economic arguments in favor of CSR and weigh in a fairly cursory way the evidence bearing on these arguments. Here I discuss an interesting possibility: CSR may provide a broader understanding of the possible economic benefits of environmental protection and worker safety than can be had from more traditional benefit–cost analysis. Finally, I express some reasons for skepticism about the CSR movement and discuss respects in which there may be less to it than first meets the eye.

Defining CSR

For the purposes of this chapter, I define CSR as *a consistent pattern, at the very least, of private firms doing more than they are required to do under applicable laws and regulations governing the environment, worker safety and health, and investments in the communities in which they operate.*[4] Thus, under this definition, a firm must consistently reduce its emissions of air and water pollution by more than it is required to do by environmental authorities to be considered as practicing CSR. It must routinely reduce risks to the safety and health of its employees to levels well below those required by law to be a practitioner of CSR. And it must invest more in those communities in which its facilities are located than it is required to do in order to be granted an operating permit—say, in the form of road building, school construction, the provision of health services or subsidies for the arts—to be considered to be engaging in CSR.

This definition is not perfect. Indeed, I will return to it later to suggest respects in which it is unsatisfactory to at least some proponents of CSR, and to me as well. But it leaves less to the imagination than definitions that make reference to sustainable development or other equally evanescent concepts. Similarly, it is much more concrete than the *description* of CSR by Holliday, Schmidheiny, and Watts, who suggest that its practitioners "focus on individuals," "instill an ethic of education," "put employees first," "keep CSR debates ... transparent," and "report externally [to] all stakeholders, not just those on their mailing list or on the Internet."[5] Although these terms are helpful in understanding what proponents of CSR are referring to, I believe my more concrete definition is more workable and also quite consistent with their ideals.

---

4. See McWilliams and Siegel, 2001, for a comparable though slightly different definition.
5. Holliday, Schmidheiny, and Watts (2002), p. 104.

There is another reason for defining CSR in the way that I have. Firms that do exactly what the law requires, but not one whit more, earn the right to do business in our society. But surely mere compliance with the law should not be sufficient by itself to earn a firm the sobriquet of "social responsibility," any more than I deserve to be considered a model citizen simply because I pay my taxes and observe laws governing drug and alcohol use. Although I will argue below that businesses do much that is extraordinarily "responsible" in the normal course of doing business and making money, most proponents of CSR are clearly focused on what we might call beyond-compliance behavior. My definition reflects their intention; in fact, it insists upon it as a minimal condition of CSR.[6]

Government Intervention in Market Economies

Most western countries, and increasingly, countries elsewhere, rely to a great extent on individuals and private firms for the provision of goods and services. There are several reasons for this, including the belief that this form of economic system is most consistent with political freedom.[7] The most common justification for the use of free markets, however, is the belief that vigorous competition among profit-maximizing firms, vying for the business of informed consumers, will result in the most efficient allocation of society's limited resources. To be clear, by the "efficient allocation of resources," I mean that it would be difficult to make some individuals better off by shuffling labor, capital or raw materials around without, in the process, making the situation of others worse. Still another way to put this is to say that given the resources available to society, markets can "squeeze out" more well-being than can systems that rely on more centralized control of economic decisionmaking.

Despite this basic faith in market organization among its proponents, virtually no one believes that the real world conforms to all the conditions required for markets to attain this ideal in the absence of any government intervention. Rather, economists and policy analysts typically identify four respects in which markets have the potential to fail, and where—conceptually, at least—government intervention has the potential to improve overall welfare. These four types of "market failure" are referred to as externalities, public goods, natural monopolies, and imperfect information. Virtually any introductory text on microeconomics discusses all of these in great length,[8] but it will be helpful to briefly illustrate each here.

Externalities are said to exist when the price of a product does not reflect all the costs that society incurs to make it available to consumers, or when at least

6. In Part I of this volume, Elhauge differentiates between corporate activity that can be justified by expected returns and that which he refers to as "profit-sacrificing"—that is, activities that will not pay for themselves. My concern is that if CSR were to be defined as only those activities that are profit-sacrificing, there would not be much to discuss.

7. Friedman (1962).

8. See Varian (1999), for instance.

some of the benefits associated with the consumption of the good or service "spill over" to third parties who pay nothing for it. Pollution is the classic negative externality. Absent government regulation, firms generally are believed to lack the same incentive to economize on the pollution they generate that they have to reduce the costly labor, capital, and raw materials they use. If the costs of pollution in an unregulated world are not reflected in the costs of production, consumers face prices that are artificially low, and too many resources flow into the production of pollution-intensive goods. Education, on the other hand, is the classic positive externality. We all benefit from a smarter citizenry, but because those investing in their own education cannot charge us for the "spillover" benefits they generate by becoming smarter (such as becoming more intelligent voters), too few resources flow into education relative to the optimum.

National defense is the definitive public good. Once an army is established to deter foreign threats, it is impossible to exclude someone claiming to derive no benefits from said protection. For that reason, free riding generally will result in the underprovision of public goods, thus again providing a rationale for government to tax all citizens and use the proceeds to provide these goods .

A natural monopoly is said to exist when the fixed costs of providing a good or service—natural gas transmission through pipelines, electricity distribution through the copper-wire grid, or cable television service, for instance—are so great that having many companies compete for customers would result in higher rather than lower costs for consumers. In a situation like this, it may make sense to have but a single provider. In this case, the government itself might provide the good or service, or, more commonly, it might grant a franchise to a single monopoly provider and then regulate the price that this firm can charge.

Finally, imperfect information can prevent the efficient allocation of resources by making it impossible for economic agents to act in their own self-interest. For instance, suppose that employers will not of their own accord reduce all serious threats to the safety and health of their employees. If workers are unaware of at least some subtle threats to their health from substances to which they are exposed on the job, they will be unable to bargain for wages that fairly compensate them for these risks. Here again, government intervention has the potential to make the allocation of resources more efficient. In the United States today, this generally takes the form of the Occupational Safety and Health Administration (OSHA) setting mandatory workplace standards limiting the concentrations of harmful substances to which workers can be exposed. In certain situations, fixing this market failure could consist simply of providing information about occupational safety and health risks to workers.

It is important to be clear that these are the principal reasons why government intervenes in the economy *on efficiency grounds*. A variety of other reasons exist for government intervention, the most obvious of which is to redistribute income. Federal tax revenue goes to the provision of public goods, of course, but it also supports income transfer programs, the provision of health care to the indigent, and other measures designed to lessen the burdens on those at the low end of the income scale. Similarly, regulatory programs sometimes are created for distributional rather than efficiency reasons. For example, during the 1970s, the government held energy prices artificially low, in part out of a concern about the impact of higher prices on the poor. Not surprisingly, efficiency and equity goals often can conflict.[9] Energy price controls, for instance, had the effects of both encouraging overconsumption of oil and natural gas and, at the same time, discouraging the search for new supplies and alternatives to them.

An additional basic point needs to be made here before I take up the cases for and against CSR. From an economic standpoint, when the government intervenes to correct market failures, it generally is guided in its search for appropriate remedies by benefit–cost analysis. Many volumes have been devoted to explaining this technique,[10] and this is not the place to review all the principles and assumptions that underlie it and its many nuances. It is, however, worth mentioning that in the application of benefit–cost analysis, a single important principle guides regulators in their design of approaches and their choice of targets.

Specifically, when they are permitted to do so by statute,[11] regulators endeavor to ensure that when pursuing environmental quality, worker safety, consumer protection, automotive fuel economy, or other social goals, standards are set at levels that equate the marginal benefits and marginal costs of additional protection. In more common parlance, they try to ensure that the last unit of protection does more good than harm. By pushing regulation to this point, but not beyond, it can be shown that resource allocation is improved—that is, that society has squeezed out the most well-being possible, given its endowment of resources. It will be important to keep this point in mind in the later discussion of CSR.

Finally, one might ask, how are the incremental benefits of regulation measured? Consider the case of an air pollutant regulated by the U.S. Environmental Protection Agency (EPA). Typically, the EPA first considers how great a reduction in emissions of the pollutant will result from the proposed controls, and then maps the reduced emissions into estimated improvements in ambient air quality. Next, using the results of toxicological, clinical, or most frequently, epidemiological studies, coupled with estimates of human exposure to ambient air, the EPA makes an estimate of the reduced incidence of premature mortality, acute and chronic illness, and other adverse health effects that could be expected to result

---

9. See Okun (1975).

10. See Gramlich (1981) and, more recently, Boardman et al. (2001).

11. Under several laws enabling regulation of the environment, worker safety, or consumer protection, no mention is made of a requirement to consider the costs of protection. The U.S. Supreme Court recently held that under these statutes, the regulatory agency is prohibited from taking costs into account in standard setting. In such cases, of course, there is no equating marginal benefits and costs; rather, the agency typically sets standards at that level likely to prevent those adverse health effects the regulator feels certain about from a scientific standpoint.

from improved air quality. If reduced levels of pollution also are expected to improve agricultural or industrial output, reduce damage to exposed materials, provide enhanced amenities, or have other favorable effects, these too are estimated.

The final step in this process, and the one that I will discuss later in the context of CSR, has to do with the valuation of these expected favorable effects.[12] If, as is generally the case, the incremental costs of air pollution control are expressed in dollar terms, so too must the incremental benefits if the "optimal" extent of regulation is to be determined. Health benefits, which generally constitute the lion's share of the benefits of environmental protection, usually are estimated by toting up the costs associated with acute or chronic illness: lost time at work, doctor and hospital bills, costs of medications, and so on. The benefits of preventing premature mortality—or the "value of a statistical life"—typically are valued by making reference to the compensation that individuals require to work in slightly riskier occupations. Avoided costs commonly are used to value materials damage and lost agricultural or forest output. In the case of benefits such as improved visibility, for which it is almost impossible to uncover evidence of individuals' willingness to pay through market transactions, surveys referred to in the literature as contingent valuation or stated-preference methods are sometimes used to elicit willingness to pay.[13]

Taken in concert, this sequence of steps, beginning with predicted changes in emissions and ending with the assignment of dollar values to types of benefit categories, results in the estimation of incremental benefits in dollar terms.

Why Engage in Beyond-Compliance Behavior?

Is government intervention required in a free-market system to force firms to internalize environmental externalities, provide adequate levels of worker safety and health, and invest in community infrastructure? Or might firms elect on their own to do more of these things than they would be required to do by law or regulation? According to proponents of CSR, firms engage in CSR for two reasons. The first, which gets heavy mention by business proponents of CSR, generally is described as a recognition by firm managers that they have some sort of moral obligation to engage in CSR-type activities because of the charter society gives them to operate. Although the nature of this obligation nowhere is defined very carefully, it is cited by CSR advocates both inside and outside of corporations.

However, another, and to me much more interesting, justification for engaging in CSR is routinely put forward: namely, that it is often in firms' economic interest to do so. In other words, proponents of CSR implicitly suggest that at least some, and perhaps a great deal of, environmental protection, worker safety,

12. See Freeman (2002) for a thorough discussion of benefit estimation.

13. See Mitchell and Carson (1989) for a comprehensive description of this technique.

and social investment would be forthcoming even in the absence of regulation. This view, too, has adherents inside and outside of corporations.

In their excellent books on corporate environmental behavior and strategy, Reinhardt and also Lyon and Maxwell go into some detail explaining the reasons why corporations may go well beyond what they are required to do by law. The highly motivated reader would do well to look carefully at both books.[14] It is worth briefly summarizing these reasons here. First, CSR proponents believe consumers will be more loyal to firms practicing CSR. This may take the form of a willingness to pay higher prices for "green" products or buy more from green firms whose prices are the same as those of other firms selling seemingly identical products.[15] Second, firms practicing CSR may find it easier to recruit skilled or highly motivated employees, or to put it differently, employees of these firms may be more productive because of the good feelings they have for their employer and the work that they do.

Third, firms that practice CSR may find their cost of capital lower than that of other firms. This will be the case if investors are more willing to purchase the stocks or hold the debt of companies committed to CSR than those of companies that are not. Fourth, firms exhibiting consistent beyond-compliance behavior may receive favorable treatment from federal or state regulatory authorities, or from the local communities in which their facilities are located. This might take the form of expedited permitting, less frequent inspections, or less local opposition to plant expansions or other facility modifications. A fifth and closely related point has to do with an even more strategic motivation for CSR. It may be the case that firms engaging in CSR are able either to preempt the more onerous regulations that they might otherwise face or even to influence the form of regulation in such a way that their competitors face higher costs than the firms practicing CSR.

The Empirical Evidence on CSR

There is a sprawling collection of studies that attempts to relate some measure of a firm's economic performance with its performance on one or more dimensions of social responsibility. The studies are by now so numerous that there exist at least 13 separate *reviews* of the literature and one very helpful book[16] that reviews 12 of these reviews, as well as the 95 separate studies that the latter collectively

14. Reinhardt (2000) and Lyon and Maxwell (2004).

15. A green firm may be one that makes a product that has a more benign effect on the environment than a competitor's product. For instance, detergent maker ABC Corporation may be seen as being a greener company than XYZ Corporation because its detergent is phosphate-free but the latter's is not. Alternatively, ABC Corporation may be seen as greener because the way in which it manufactures its detergent is less harmful to the environment than XYZ Corporation's method, even though the characteristics of their two products are identical.

16. Margolis and Walsh (2001).

comprise. In addition, the books by Reinhardt and also Lyon and Maxwell,[17] as well as the chapter by Reinhardt in this volume, also summarize in a nice way many of the important empirical studies to date. None of the individual studies derive testable hypotheses from a theoretical model of the firm, however, and few of them are very clear about the specific mechanisms through which firms' socially responsible behavior is supposed to work to their financial advantage.[18]

Consider first the possibility that firms may be able to differentiate their products on the basis of the firms' environmental, occupational, or social investments or actions, and thus enjoy some pricing power relative to their price-taking rivals. Vignettes abound on the subject of green consumerism. Cairncross compiles the best collection of examples seeming to show that environmentally sensitive products, or products produced by companies using green production techniques, have been greeted especially warmly by consumers.[19] Yet the literature on CSR contains no study that attempts to measure price–cost margins for a standard product or set of products and relate them, in a careful multivariate analysis, to one or more measures of CSR, however measured.

In addition, though perhaps less numerous than the examples Cairncross presents, notable examples of failed green marketing strategies exist. For instance, the former gasoline marketer ARCO introduced a reformulated and significantly cleaner gasoline in Los Angeles in the early 1990s, well before all companies were required to do so. Moreover, this introduction was preceded by a careful consumer marketing survey, in which people were quizzed about their willingness to pay for cleaner fuels, and was accompanied by a sophisticated public information campaign featuring spokespersons from the environmental and public interest community who were believed to have great credibility. Yet as one ARCO executive said at the time, "People drove past our gas stations in droves to those of our nearby competitors where gas was selling for $0.05/gallon less."[20] If green consumerism is one route through which CSR is believed to pay off for firms, careful analysis would make a welcome contribution.

As suggested above, a second route through which CSR could pay off in enhanced profitability relates to a firm's employees. Indeed, this seems to be the most plausible mechanism through which CSR might be financially prudent, as virtually everyone can imagine employers for whom he would not work and others with whom he would be proud to associate. However, the CSR literature con-

---

17. Reinhardt (2000), Lyon and Maxwell (2004).

18. A theoretical model of a firm engaged in CSR activities would not be difficult to sketch out. In such a model, the price charged by the firm, as well as the wage rate and the cost of capital it faces, would not be exogenous to the firm. Rather, they would depend in part on a set of environmental activities in which the firm is engaged. Thus profit maximization would depend not only on the cost of engaging in these environmental activities, but also on the favorable effects these activities would have on the price the firm could charge and on the wage rate and the rental price of capital it faced.

19. Cairncross (1992), especially pp. 189–211. See also McWilliams and Siegel (2000) for an endorsement of green consumerism.

20. Personal communication with author.

tains no studies that attempt to link beyond-compliance activities to either lower wages or enhanced productivity, at either the firm or industry level. To be sure, a vast literature shows that differences in the injury or fatality rates associated with various *occupations* imply differences in compensation, controlling for other factors that affect the wage rate.[21] Thus firms that spend more money than they have to in order to make their workplaces safer should expect to pay lower wage rates, or have more productive workers, than those that do not.

Moreover, anecdotes abound about firms that have worked to improve their image so that they might continue to attract outstanding new employees. Dow Chemical Company talks openly about its quite real environmental "makeover" in the 1980s and early 1990s, prompted in part by an interest in the continued successful recruitment of outstanding young engineers. But I could find no study that looked at differences in firm- or plant-level wages or productivity and attempted to explain them using measures of the CSR differences between firms, again while controlling for other possible influences.

What about differences in the cost of capital among firms? Might CSR activities enlarge the pool of investors a firm might attract, or might investors be willing to accept lower rates of return from firms seen as behaving in a socially responsible way? Once again, the evidence is scantier and less conclusive than one would prefer.[22]

There can be no doubt that investments in all assets managed with at least one eye toward social and environmental consequences have grown dramatically. According to the Social Investment Forum, such assets amounted to more than $2.2 trillion in 2003, about 11 percent of all professionally managed assets. During the period 1995 to 2003, assets professionally managed using one or more CSR "screens" grew by 240 percent, while all professionally managed assets grew by 170 percent. Mutual funds alone that incorporate CSR screens—which numbered 200—contained $162 billion in 2003.[23]

The screens to which analysts or proponents refer when discussing CSR funds are not restricted to the environment, worker safety, and community investment, although these figure prominently in the stock selection process of many of the socially responsible funds. Briefly, looking over the various screens that are commonly used, firms wishing to pass muster with those managing such funds should eschew lines of business that involve alcohol, tobacco, gambling, nuclear power, firearms, defense contracting, abortion or birth control, pornography, money lending, and animal testing. They also should avoid investments in Burma, China, and other countries in which CSR proponents believe indigenous peoples are mistreated. Rather, firms wishing to attract the support of socially responsible investors should embrace shareholder advocacy; research and development; renewable energy; strong union relationships and employee empowerment;

21. See Viscusi and Aldy (2003).

22. See the very helpful essay by Lloyd Kurtz that can be found at http://sristudies.org/essay_frameset.html (accessed September 1, 2004).

23. Social Investment Forum (2003).

and the recruitment of employees, managers, and directors from a broad spectrum of people with diverse racial, gender, religious and ethnic characteristics, sexual orientations, and attitudes.[24]

It should go without saying, and proponents of socially managed funds will readily admit, that ambiguity exists in such criteria. For instance, medical research suggests that alcohol in the form of a glass of wine each day is good for one's health. Similarly, although nuclear power surely presents formidable problems regarding the storage of spent wastes and the possible proliferation of nuclear weapons, electricity generation from commercial nuclear reactors is free of such conventional air pollutants as sulfur and nitrogen oxides, hydrocarbons and particulate matter, and—perhaps most important—carbon dioxide, the predominant greenhouse gas. Finally, one need not live near the Pentagon or White House, as I do, nor reside in the vicinity of what was once the World Trade Center, to believe that firms doing business with the Department of Defense are protecting the United States and its citizens, *not* engaging in activities that are socially irresponsible.

The merits of screens for social responsibility aside, how well do socially managed funds do when compared with more traditionally managed mutual funds? Here, too, the evidence is difficult to encapsulate neatly, depending as it does on which funds are compared with which, and for what periods of time. Using information provided by the Social Investment Forum, Stone et al. found in 2001 that over the period 1984 to 1997, the introduction of social screens did not hurt the returns of stocks selected by a traditional model.[25] Also, Bauer, Koedijk, and Otten found that the risk-adjusted returns of more than 100 U.S., German, and UK firms did not differ when they were broken down between those that used social screens and those that did not.[26]

More recent evidence, however, seems to suggest that socially screened funds have slightly underperformed traditional mutual funds, and for a subtle reason. Recall from the list of screens above that investments in R&D are seen as a good thing when screening for green companies. So, too, are the absence of pollution problems and the presence of stock ownership plans for employees. These commonly are characteristics of so-called high-tech firms, especially those organized around the Internet and other communications and information technologies. Such firms have invested heavily in R&D; generally have few pollution problems, often because they do little manufacturing or contract it out to other manufacturers and merely assemble the parts they purchase; and frequently have granted stock options to new employees. This meant that many socially managed funds tended to be "tech-heavy"; for that reason, they could have been expected to do better than more traditionally managed funds until the beginning of the NAS-

---

24. Geczy, Stambaugh, and Levin (2003), pp. 26–27.
25. Stone, Guerard, Gutelkin, and Adams (forthcoming).
26. Bauer, Koedijk, and Otten (2002).

DAQ decline several years ago, and to do worse once the more recent period is factored into the comparison.

In one of the very few theory-based research papers in the CSR literature, Geczy, Stambaugh, and Levin ask an interesting question: supposing that socially motivated investors are willing to accept lower returns from CSR funds, how big a penalty must they pay? That is, how would the returns from an optimally constructed set of CSR funds differ from those chosen from the much larger universe of all mutual funds? They find that the answer depends on investors' views about how assets are priced (whether they believe in the Capital Asset Pricing Model or Fama's and French's three-factor model, for instance), their views concerning the skill of fund managers, the share of their total portfolio devoted to CSR funds, and the severity of the screen used to differentiate among firms. Depending on these beliefs, CSR investors could pay a financial penalty of as little as 1 or 2 basis points per month (in other words, their returns would be less than those of "unconstrained" investors by one or two one-hundredths of 1 percent each month), or as much as 1,500 basis points monthly (earning 15 percent less per month).[27]

To summarize, when one looks for evidence on the *specific mechanisms* through which investments in CSR ought to benefit the firms that make them, it is scant. Beyond a series of sometimes compelling examples of companies that have been able to differentiate their products, and hence attract the attention of green consumers, one cannot point to any thoroughly systematic evidence or serious empirical work. The same is true of a possible link between investments in CSR and a more productive or compliant employee base. Although it seems obvious that employees who are well treated will be better for an employer than those who are not, this relationship has yet to be tested rigorously.

Finally, there can be little doubt about the growth of assets that are managed at least in part with an eye toward CSR, as determined in a wide variety of ways. To the extent that these funds earn better returns than their conventionally managed competitors, they no doubt will attract wider and wider support—and, in the process, provide a boost to the companies that they comprise. To this point, however, no convincing evidence exists that such funds have earned consistently superior returns. If this continues to be the case, investors will choose socially managed funds with their hearts rather than their pocketbooks; this suggests that firms practicing CSR will be unlikely to face lower capital costs than other firms.

This does not mean, however, that evidence is lacking altogether on a possible link between a firm's environmental or social performance and its financial record. Indeed, although the studies that it comprises are silent on just how the mechanism works, a large literature, including a number of statistical analyses, attempts to explain the financial performance of a large set of companies by, among other things, its commitment to CSR.

27. Geczy, Stambaugh, and Levin (2003).

It is impossible even to begin to describe this literature here. For a comprehensive overview, the reader should see Margolis and Walsh (2001). Suffice it to say that the 95 studies they survey measure firms' economic performance in a variety of ways, including their cumulative abnormal return, as well as their return on equity, assets, and sales. An even larger and more diverse set of variables is used to measure some dimension of firms' commitments to CSR. These include some measures related to the firms' environmental performance; the products or services they sell; their corporate governance practices, including such things as the presence and power of outside directors; and their investments in countries that are considered to have intolerable political or labor practices. In the statistical analyses that go beyond identifying simple correlation coefficients between CSR and financial performance, other independent variables frequently include some control for the industry in which the firm operates, the size of the firm, its debt-to-equity ratio, and the intensity of its R&D and advertising.

Margolis and Walsh conclude that in slightly more than half (53 percent) of the studies using CSR as an independent variable to explain financial performance, a positive relationship is found (they do not mention whether these relationships are statistically significant). In a quarter of these studies, no relationship is found, and a negative or mixed relationship exists for the remaining quarter.

It is essential not to read too much into any summary, no matter how thoughtful, about the literature attempting to link CSR to financial performance. As almost all the summarizers point out, measurement of the dependent and independent variables in the studies is fraught with difficulty. Much more importantly, the specification of the estimating equations in these studies is often simplistic. In some cases, for example, a firm's commitment to CSR and its financial performance are measured contemporaneously, making it impossible to determine the direction in which any possible causality may run.

In addition, omitted variable bias greatly confounds efforts in some of the studies to draw statistically valid inferences about the causal effects of CSR. McWilliams and Siegel reestimate a model used in one study, in which the independent variables being used to explain firm profitability include a measure of the firm's CSR efforts, its size and debt–asset ratio, and an industry dummy.[28] When the estimating equation is rerun with the addition of a measure of R&D intensity (the ratio of a firm's R&D expenditures to its sales), the coefficient of the CSR variable falls in size and becomes statistically insignificant. Such findings might become common if many of the empirical studies were redone in an econometrically more rigorous way.

A Normative Perspective on CSR

Whether firms *do* go above and beyond what they are required to do by law or regulation—that is, whether they engage in CSR, using the definition presented

28. Waddock and Graves (1997).

earlier in this chapter—seems beyond doubt. The strongest proponents of CSR,[29] as well as its most vigorous critics,[30] cite example after example of corporate commitments to a wide variety of activities and investments that firms are not required to make. To this point, the discussion has raised and attempted to provide at least rudimentary answers to two important questions: Why might firms go beyond what they are required to do in such areas as environmental protection, worker safety, and health and community development? And is there evidence to suggest that doing so is to their financial advantage?

These are not the only interesting questions one can ask about CSR. Speaking normatively, for instance, how *should* we feel about such activities on the part of firms? That is, should they be encouraged, tolerated, or frowned upon?

These may seem like strange questions to ask. It seems natural to think about constraining the activities of corporations where they involve the creation of negative externalities or monopolies, for example, but why would we want to discourage their pursuit of apparent good works? Indeed, David Henderson, by far the strongest and most articulate critic of the CSR movement, acknowledges that it is odd for outsiders to question the decisions made by firm managers, especially when "their shareholders appear to have accepted the various moves to CSR with equanimity or even approval."[31] Presumably, these managers are in the very best position to know what is in the interests of their firms and what is not.

Still, it is worth asking whether and under what conditions firms' CSR activities could improve the current allocation of resources—worth asking, that is, whether these activities *could* be efficient in the economic sense. When the question is posed this way, the answer is, indisputably, yes.

For example, no one would argue that every serious threat to consumers, workers, or the environment has been addressed, even in the western industrial democracies where regulation is pervasive. Regulatory authorities often are given or take on much more work than they have resources to deal with. For that reason, serious potential problems often go unaddressed. In the environmental area alone, despite 32 years of significant federal regulation of water pollution in the United States, we still lack meaningful binding constraints on so-called "nonpoint" sources of water pollution, including agriculture, municipal runoff from storm sewers, and coastal development. Similarly, although large animal feedlots have become significant sources of air and water pollution in many parts of the country, only in the last several years have regulations come to be developed for them. More broadly, in some rapidly growing parts of the developing world, almost no regulatory authorities may be in place, even though environmental, occupational, and consumer risks may be greater there than anywhere else in the world.

---

29. Holliday, Schmidheiny, and Watts (2002).
30. Henderson (2001).
31. Henderson (2001), p. 66.

In cases where governments have yet to regulate a potentially serious risk, and especially in those cases where that risk can be inexpensively controlled, it is likely that the marginal social benefits of CSR actions by the firm to lessen the risk will exceed the incremental costs. Although he does not couch his discussion in welfare-theoretic terms, Henderson seems to acknowledge this point. He writes, "Both shareholders and boards of directors may be willing, and arguably should be willing, to risk or forgo profits at the margin for such causes as ensuring product safety, disclosing possible safety risks [and] reducing harmful pollution ... even where no legal obligations are in question."[32] With the benefit of hindsight, had the companies that produced and made use of asbestos in their operations in the United States many years ago acted on the health information they appear to have had in their private records and reduced worker exposure, they would have avoided health costs and legal liabilities that surely exceeded their savings at the time.

It is likely that one type of investment in CSR will be especially efficient. When multinational firms build a new plant in developing countries, it is generally the same or very similar to the one they would erect in a developed country.[33] This is the case even if environmental and workplace health standards are significantly less stringent, even nonexistent or largely unenforced in the developing country. This makes economic sense from the firm's standpoint, because building a single, state-of-the-art plant wherever they are locating may cost much less than tailoring plant designs to the special standards of many different possible host countries. The key here is that health and environmental protections can be offered for little additional cost. For reasons to be discussed below, one should *not* jump to the conclusion that it generally will be efficient for western companies to offer all of the same protections and benefits in the developing world that they do at home.

Corporate investments in CSR often take the form of contributions to the educational systems of the communities, states, or even countries in which they have operations. This may involve building schools, donating books or supplies, lending their employees as part-time teachers, or providing scholarships. As with government investments in education, because of the positive spillovers thought to be associated with education, CSR investments by firms easily can generate benefits in excess of their costs. So, too, could such investments in health care as free vaccination programs, as positive spillovers have been shown to be significant for such initiatives. To recap, CSR investments, especially those that are relatively inexpensive for a firm to make, likely will generate positive net benefits in many instances.

CSR activities might be economically efficient in another respect, even from the narrow standpoint of the firm making such investments, and even if consumers, workers, or investors paid no attention to the firm's efforts. It has to do

32. Henderson (2001), p. 22.

33. See Jaffe, Peterson, Portney, and Stavins (1995).

with the idea of social norms.[34] The idea here is that firms might take actions to reduce their pollution, say, if they realize the damage being done by the pollution is greater than what they would have to spend to clean it up. They might do so not because they think their action will be rewarded by consumers, workers, or investors, but because they believe that everyone would be better off in the long run if every firm behaved that way. By analogy, each of us would go to some effort to return a lost wallet that we found. We do this not because the person to whom we return the wallet might one day find ours (the chances of which are infinitesimally small), but because if everyone behaves in this way, we are highly likely to get our own lost wallet back one day.[35]

It is useful to go beyond these examples of specific types of CSR investments that may improve the allocation of resources to a more general, though controversial, point. The most interesting possibility regarding investments in CSR is that they could prove to be a better barometer of benefits than could be had through some traditional benefit–cost analyses. This is a complicated argument to make, but it proceeds in the following way.

Suppose, first of all, that firms practice CSR because they believe it is in their financial interests to do so. That is, they engage in beyond-compliance behavior with respect to the environment, worker safety, and social spending because they believe that the financial returns they would reap from these actions are comparable to the returns from more traditional investments such as research and development, the construction of a new plant, or the development of a new product. In the case of CSR investments, the returns take the form of customers willing to pay a higher price for their product, workers who are more productive, or investors who are more willing to purchase the firm's equity or debt.

Consider next just such a firm—one that is considering enhancing the safety of its manufacturing facilities far beyond what it is required to do under current occupational safety and health standards. To make the matter concrete, suppose it is contemplating the complete elimination of an airborne contaminant found in its factories that may pose some risk to its workers.

In making a calculation whether to undertake this investment, the firm would consider the effects such an action would have on its workforce. That is, the firm would attempt to determine whether its workers would be healthier, and thus more productive, if it eliminated the contaminant; whether they would be more productive because they felt that they worked for a benevolent employer; and whether such an investment would reduce any possible liability for its employees' health damages in future years.

---

34. I thank Jon Sonstelie for suggesting this point to me, which seems related to Dan Esty's interesting suggestion later in this volume about CSR being a kind of "winner's tax" on firms that have done well in their respective businesses.

35. True enough, once a social norm has been established, it becomes possible for a free rider to evade his "finder's" responsibility and still be confident that he will get his wallet back from someone else.

The firm would not stop in its calculations there, however. It would ask as well whether such an action, if made known widely,[36] might also enable it to attract customers who approve of such behavior on its part, and whether this change might make the firm more attractive to socially minded investors. If the firm is correct in its assessment, and if customers and investors really are willing to reward such behavior, that means that the latter are willing to pay for an improvement in workplace safety even though that change would do nothing to improve their own health.

Contrast this with the benefit–cost analysis likely to be conducted by OSHA. Typically, when analyzing a proposed regulation, OSHA will make an estimate of the number of workers likely to benefit from a reduction in a workplace hazard, usually based on extrapolations to humans from animal toxicological tests or on epidemiological studies. It then would attach a dollar value to each case of illness or injury prevented and count the resulting product as a measure of the incremental benefits, to be compared with incremental costs when and if it is allowed to do so by law. It would be most unusual for OSHA to attribute benefits to those outside the workplace—consumers or investors—who take some satisfaction from knowing that workers are safer there.[37] To the extent that this latter sentiment is real, and to the extent that the firm considering the CSR investment is correct that it will be rewarded for its voluntary action, it has made a more accurate determination of benefits than OSHA would.

Other hypothetical examples abound. When the Consumer Product Safety Commission (CPSC) sets safety standards for cribs, it determines how many fewer infant accidents or deaths will result, attaches a value to preventing such incidents, and comes up with an estimate of incremental benefits. Suppose, however, a crib maker voluntarily takes additional safety measures not only so that it will sell more of the safer cribs, but also because its workers will be proud of its actions and thus more productive, and because the company will be more attractive to investors. It may be registering incremental benefits and equating them with costs more accurately than CPSC would in its regulatory analysis, so long as workers and investors really are willing to put their money where their social conscience is.

For those familiar with the nuances of benefit–cost analysis, especially as it is applied to environmental issues, this concept will be familiar. What is being sug-

---

36. It is common for corporations today to issue not only annual financial reports, but also reports on their environmental and social performance. In these reports, which are sent to shareholders as well as to other interested parties, companies discuss in some detail their investments in CSR activities, often listing the nongovernmental organizations whose activities they support. Additionally, these reports frequently show voluntary reductions in pollutant emissions, changes in safety performance, and interactions with various "stakeholder" groups. Companies sometimes advertise widely in print and broadcast media, as well, chronicling their CSR actions.

37. I am aware of no case where OSHA has taken into account in a benefits analysis any such effects.

gested here is that under certain conditions, voluntary CSR activities by firms may lead to the inclusion of *existence values* that would be missed in a traditional benefit–cost study. Such values arise when parties who appear not to be directly affected by a regulatory action nevertheless feel a real stake in it *and would be willing to pay something to make it happen.*

It is often difficult in benefit–cost studies to assign dollar values even to the improvement in well-being felt by those directly affected. For instance, how does one assign a value to health improvements experienced by small children who face reduced risks from, say, air pollution or exposure to hazardous substances on the playground? How should the value of odor reductions from animal feeding operations be calculated for those who live nearby? How might we express in dollar terms the value to aspiring parents of a reduced risk of infertility associated with the elimination of possibly hormone-disrupting chemicals in the environment?

As daunting as these or comparable challenges might be, they pale in comparison to ascertaining the value, say, of knowing that a pristine wilderness area exists even though one never intends to visit it, or of knowing that someone else's workplace is safer. It is for this purpose that the contingent valuation or stated-preference technique was developed, and it has been controversial almost ever since.[38] What is being suggested here is that consumers, workers, and investors concerned about CSR-type issues may have a "market" in which to express their preferences for a cleaner environment, safer workplaces, or other good things, and thereby provide information unlikely to arise in a standard benefit–cost analysis. If so, it is at least conceptually possible that CSR could result in a better balancing of benefits and costs than regulators could accomplish.

Symmetry requires us to ask whether circumstances exist in which investments in CSR *could* be welfare-reducing. Here, too, the answer is yes. In fact, in his elegant critique of it, Henderson claims, "CSR rests on a mistaken view of issues and events, and its general adoption by business would reduce welfare and undermine the market economy."[39] Although it is easy to be sympathetic to many of the points Henderson makes, there is reason to be far less certain and somewhat more sanguine about what effect CSR *will* have. But it is not difficult to point out respects in which it could cause mischief.

The principal concern about CSR in general is that it will divert private corporations from the purpose for which they were created—to provide goods and services to consumers under the protection of limited liability. Firms making CSR investments that are not expected to earn competitive returns are diverting capital from more lucrative opportunities; executives or managers devoting time to the many soirees on CSR that are constantly being held could be thinking about ways to make their product better or reduce the cost of producing it. In other

38. See Diamond and Hausman (1994), Hanemann (1994), and Portney (1994).
39. Henderson (2001), p. 18.

words, CSR has an opportunity cost, as does everything else. However, if firms' commitments to CSR are confined mainly to enhanced annual reporting, meeting periodically with and listening to stakeholders, making more transparent their internal governance structures, and other relatively simple things, these commitments will not be the firms' undoing and will not greatly distort the allocation of resources.

Moving to a much greater level of complexity, there are other respects in which CSR activities could be welfare-reducing rather than welfare-enhancing.[40] These arise from models in which firms act strategically to anticipate future regulations and shape their form. For instance, firms perceived as being leaders in their industries might voluntarily adopt standards that signal to regulators that regulation is more expensive than the latter believed it to be. To the extent that this leads the regulator to set less stringent standards than it would have otherwise, welfare can be diminished rather than increased through CSR.

One possible interpretation of CSR has the potential to be especially harmful, though in a way that is subtle to those unused to thinking about international trade. Some advocates of CSR, though generally not those within business, push firms toward the harmonization of environmental and labor standards across international boundaries. Thus proponents of "fair trade" suggest that companies manufacturing in both the developed and the developing world should offer comparable wages and benefits wherever they produce. This same line of reasoning is used to support the harmonization of environmental and workplace safety standards across countries as well.

Although well-intentioned, this call overlooks the fact that differences in incomes across countries and over time give rise to very different demands for, and hence valuations of, environmental protection, workplace safety, and other objects of CSR. For instance, Americans demanded less environmental quality 50 years ago but have come to insist on more and more of it as their incomes have grown and their more basic needs have come to be met. By this logic, it is inappropriate to expect those in the developing world, where incomes are anywhere from one-tenth to one one-hundredth of those in the United States or western Europe, to demand the same standards that we in the developed world enjoy today. Incorrectly assuming that they do would lead firms to spend too much on environmental and other protections (strange as that may seem to some) and too little on wages and new job creation. In the process, firms would significantly reduce the well-being of those very people whom CSR proponents claim to want to help. The case of China, which has moved to dramatically reduce its air pollution, shows very clearly that incomes do not have to rise to anything approaching western levels before people begin to insist that at least part of their increasing standard of living take the form of a cleaner environment or improved working conditions.[41]

40. This point is developed quite elegantly using formal theoretical models of both firm and government decisionmaking in Lyon and Maxwell (2004).

41. Interestingly, China is on the brink of adopting standards for automobile and light-duty truck fuel economy that will be significantly *more* stringent than those in the United States.

To summarize, if the pursuit of CSR leads firms to control pollution, improve the safety of their workplaces, redesign unsafe products, or make strategic investments in education, health, or community infrastructure, it is very possible that these actions could be welfare-enhancing. This is more likely to be the case if the firm is pursuing CSR strategies because it believes they will boost its financial bottom line, and more likely still if such initiatives are relatively inexpensive. On the other hand, if firms fully embrace the more ambitious versions of CSR—especially the view that they should pay well above prevailing market wages in developing countries and embrace the same environmental and workplace standards there that they face in the developed world—welfare almost surely will be reduced.

Three Concerns and a Conjecture about CSR

The preceding discussion brings us back around to a broader view of CSR, in which three distinct concerns arise. The first is a semantic one, but an important one nevertheless. Proponents of CSR are clear in insisting that in order to be considered socially responsible, a firm must engage in a variety of activities that take it well beyond those in which it would engage in the normal course of profit maximization. This is pernicious. It ignores the fact that in the normal course of attempting to make money, firms both large and small routinely do a number of things that are extraordinarily "responsible" and that should not be taken for granted.

For example, in 2002, more than 108 million people were employed in the nonfarm private sector in the United States, almost 20 million more than just 10 years ago. Many of these workers received as conditions of their employment not only paychecks, but also contributions toward their and their families' health insurance and social security, and in many cases, toward additional retirement accounts or plans. Given the importance of work in many people's lives, not to mention their need to make a living, it seems strange to relegate this accomplishment to the realm of the ordinary simply because private employers hire people to help them make money.

Private businesses do other valuable things, of course. For instance, corporate debt in the United States currently stands at about $5 trillion, $20 billion of which changes hands daily. Fifty million Americans now own stock, including $2 trillion in equity in manufacturing corporations alone. Thus in addition to the jobs it creates, the private sector offers a ready outlet for savings, encouraging both thrift and risk-taking, qualities that are highly prized. Finally, businesses sell in excess of $7 trillion in goods and services each year, all of which presumably were worth more to those who purchased them than they had to pay. At the risk of sounding dazzled by Adam Smith's invisible hand, we should not let the fact that the private sector does all these things to earn a competitive return dull us to their importance.[42]

---

42. This presumes, of course, that firms pursue profit maximization in an entirely legal way. It should go without saying that one key precondition for a properly operating system of essentially free markets is a legal regime that is fairly and vigorously enforced. One can

A second concern about CSR is definitional. The argument immediately above notwithstanding, virtually everyone weighing in on CSR takes it to mean the set of actions in which a firm engages beyond those required by law or regulation—thus the definition put forth at the beginning of this chapter. But a careful reading of the contributions to the literature on CSR from those in the corporate world makes it clear that they engage in beyond-compliance behavior because they feel that most or all of these actions pay off economically. Virtually all of the 67 case studies presented in the book by Holliday, Schmidheiny, and Watts (2002) on corporate engagement in CSR make reference to or provide estimates of the money saved through a wide variety of CSR initiatives.

This is fine and good. As argued above, many good things happen when companies set about to make money, even when they do no more than the law requires of them. And more of these good things will follow if and when companies are correct in taking an even broader view of what they might do profitably. However, if companies engage in environmental good works, make their workplaces safer than required, and invest in the communities in which they are located because these investments have an internal rate of return greater than the firm's "hurdle rate" for more conventional investments, two questions are unavoidable. How is this behavior any different from "ordinary" profit maximization as might be practiced by firms seen to be intransigent by CSR proponents? And if a firm's investments in CSR are defended on the grounds of profitability, what is so socially responsible about that?

It is perfectly understandable why firms would like to get credit in the form of goodwill for their CSR investments, over and above the money they save or the profits they reap from these investments. Indeed, even firms eschewing CSR regularly tout the number of people they employ locally, the amount of money they inject into local economies, the taxes they pay, and so on—even though they do these things to earn profits. There is no reason why firms going beyond what they are required to do should not also strive to be seen as good citizens for this. In fact, getting public acclaim for their CSR activities may bring them even more customers and investors and make their workers even happier. But it is hard to see why such firms should be seen as any more virtuous—more socially responsible, that is—than firms that have a narrower conception of how they might prosper.

Consider for a moment those companies—call them true believers, if you will—who pursue not only those CSR investments that they believe will pay themselves off, but even some they strongly suspect will not.[43] A few brief observations are in order. Unlike Henderson, I do not believe that such investments are all that common or that true believers are very numerous. As domestic markets become ever more competitive, and competition from foreign sources

---

admire the marvels of the private-enterprise system while still being contemptuous of those who would imperil it by flaunting the laws and regulations that make it possible.

43. These are the activities that Elhauge refers to as "profit-sacrificing" in Part I of this volume.

increases, cost cutting is a high priority throughout the corporate world. It is unlikely that many companies are making sweeping commitments of time or money to good causes they do not believe will pay off in some way for their companies. Also, even if this behavior is as common as Henderson fears, it is not likely to persist for long if markets are highly competitive and if the CSR investments in question are more than tokens. For if the latter is the case, the goods or services offered by true believers will become more expensive than those of competitors, and a Darwinian culling out could be expected.

A third observation about CSR is premised on the notion that a particular firm is a true believer and invests in activities that hold no promise of earning a competitive return. This may be because the firm is privately owned rather than publicly traded. In this case, it should feel free to spend its money however it sees fit, in the same way individuals are free to do so. In fact, it would make an interesting research project to see whether privately held companies (which include such prominent corporations as S.C. Johnson & Son, Mars, Cargill, and Koch Industries) are more avid practitioners of CSR than their publicly traded counterparts.

Suppose, though, that the firm is publicly traded and is able to engage in profit-sacrificing activities because it is protected from the forces of competition. In this case, it is not clear how one ought to regard its CSR activities. The money its managers spend on CSR activities belongs to its shareholders. This money could be paid out to shareholders in the form of higher dividends, or it could be retained and reinvested in the profit-making parts of the business, in which case it would also redound to the benefit of the firm's shareholders.

If firm managers feel that a particular cause is a good and just one, they could contribute their own time and own money to it. Many no doubt do just this. They also could spend time trying to enlist the support of others in this cause, including local, national, and international charities or lending organizations. They might even call the cause to the attention of their shareholders in the hope that the latter, too, will be moved to support it. But it is not at all clear that they ought to devote large sums of their shareholders' money to even the worthiest of causes; a strong case can be made that such philanthropic decisions are better left to shareholders.

This brings us, finally, to a brief conjecture about the CSR movement, specifically the possible motivations behind it. Why is it that firms in the private sector, and principally large corporations, are being pushed to spend their money on environmental enhancement and worker safety beyond the dictates of prevailing standards? Why are they asked to build schools, donate equipment and teachers, and do other things to enhance public education? Why should profit-making corporations purchase or donate vaccines for the poor or build health clinics to serve them? And why should they be the underwriters for the ballet, the opera, the symphony, and the art museums?

Proponents of CSR have many explanations to offer, at least some having to do with a moral obligation they feel corporations have as part and parcel of their public charter. But there are two possible explanations for the current enthusiasm for CSR that are worth noting, both of which have important implications for economic efficiency and public policy.

Some enthusiasts for CSR may be motivated by a lack of confidence in the institutions of government or an unwillingness to fight for the higher taxes that would be necessary for the public sector to undertake activities that are legitimately in its domain.[44] Tax increases of almost any sort apparently have become anathema to almost every politician. Accordingly, the federal budget is in deficit once again, as are the budgets of many state and local governments. The response to these deficits is almost always the same: cut spending. Similarly, there appears to be less confidence in the ability of legislatures to enact new regulatory statutes and of regulatory agencies to issue the new rules that proponents feel are needed. Thus some who have turned to CSR appear to be motivated by the belief that it is only through private-sector spending that some desired activities will be possible.

But this can be viewed differently. If the public is unwilling to pay higher taxes for better schools, new public health initiatives, enhanced infrastructure, or a more robust arts program, and indicates such when these programs are put to votes, it is saying that, at the margin, at least, these programs are not worth to them what they would have to pay in higher taxes. In other words, the perceived benefits of these public goods fall short of costs. Making such goods available through corporate provision would not change this fact, though it would change who pays for them. This leads directly to the next point.

A second possible reason for the growth of enthusiasm in CSR has to do with a kind of "fiscal illusion." Some proponents of CSR may have the view that if companies are talked or coerced into providing services that are the legitimate province of the public sector, these services become free. Though no one comes right out and says this, it is not uncommon to hear some people say, "Corporations should pay for this." But corporations are disembodied entities, mere creatures of law, and if they provide what are essentially public goods, the public pays for these goods in one form or another. This may be in the form of higher product prices, if the corporation has pricing power, or it may take the form of reduced job opportunities or lower earnings for shareholders, if the costs cannot be passed forward. In any event, real people will bear the burdens, not some legal construct called a corporation.

Whether out of frustration with a withering public sector or in the misguided belief in a free lunch, pushing corporations to provide public goods has important distributional implications. They cannot be spelled out with any specificity. The precise distributional consequences depend upon how the cost of public goods would be financed by the public sector—through income or property

44. As indicated above, education, certain public-health measures, and a clean environment are classic public goods and therefore appropriate activities for collective action. Support for the arts has always been a harder call; although high culture has some public-good elements, it generally is possible to exclude those who wish not to support it. Public support for the arts usually is justified more on the grounds of making them available to those at the lower end of the income scale.

taxes, user charges, or excise taxes, for instance—and how they would be paid for if they were pushed onto the private sector, through higher prices, reduced wages or job opportunities, or reduced shareholder earnings. There is no denying, however, that different groups will pay in the two cases, and that it is important to understand the differences.

Closing Thoughts

Because CSR is an inherently fuzzy construct, it is not easy to analyze from an efficiency or public policy perspective. Nevertheless, a number of points are worth reemphasizing here. First, if it is to mean anything, CSR must be construed as the set of activities in which private firms engage beyond those that they are required by law to undertake. Even this definition becomes somewhat problematical if firms go beyond compliance because doing so is profitable for them, but at a minimum such behavior ought to be required under a reasonable definition of CSR.

CSR activities might redound to the economic benefit of its practitioners through a number of mechanisms, and a large and growing literature is devoted to testing this link empirically. Unfortunately, many of these studies (though by no means all) are less rigorous than one would like, and they point in no consistent direction. If firms are looking for solid and conclusive evidence that their bottom lines will improve if they devote themselves to CSR, they will be disappointed.

It is more difficult still to say whether firms should be encouraged to undertake investments in CSR. This will depend upon both the nature of the investment and its size. There surely will be cases when CSR investments will be welfare-enhancing—that is, when they will improve the allocation of resources in the economy. This is most likely to be the case when firms act to control pollutants or workplace hazards they know to be serious and that are currently uncontrolled; such actions are even more likely to be desirable from society's standpoint when they are relatively inexpensive. By the same token, one can easily imagine CSR activities that would do more harm than good, the paradigmatic example being efforts to pay uniform wages or enforce the same environmental standards around the globe.

As a practical matter, the likelihood of great inefficiencies resulting from CSR seems small. Firms are under constant pressure to reduce costs and much more often than not will go beyond what they are required to do only when it is profitable for them to do so. Their forays into pure and unadulterated philanthropy are likely to be few and relatively minor. Appropriate public policy regarding CSR is that of laissez-faire. Firms will know best when their investments will pay off and when they will not, and it would be strange to caution them against good works on the grounds that they will go broke engaging in them.

A more troubling aspect of the CSR debate is that it cheapens the many quite valuable things that firms do, including providing livelihoods to people, creating an outlet for their savings, and producing an unbelievably wide variety of goods and services that are eagerly snapped up by consumers. To be clear, firms do *not* do these things out of the goodness of their hearts, but to make money, purely and simply. Nevertheless, their profit-driven activities seem infinitely more valuable than the things firms do under the CSR umbrella to curry favor with their customers, employees, and investors, and with their critics in the social activist community.

Finally, those pushing CSR—from outside the corporate world, at least—may be doing so in part because it is so hard to accomplish things through the public sector. Another way of saying this, though, is that if the public is not willing to tax itself to have these benefits, they may not be worth as much to the public as proponents would believe. For that reason alone, it may not make sense to foist these activities off on the private sector, even if one could do so successfully.

Acknowledgments

For helpful comments on an earlier draft, I want to thank Dennis Aigner, Daniel Esty, David Henderson, Dale Heydlauff, Gerald Keim, Virginia Kromm, Richard Liroff, Thomas Lyon, Ian Parry, Donald Siegel, Jon Sonstelie, Robert Stavins, John Tilton, and Margaret Walls. Katrina Jessoe provided excellent research assistance, for which I am grateful.

References

Bauer, Rob, Kees Koedijk, and Roger Otten. 2002. International Evidence on Ethical Mutual Fund Performance and Investment Style. Limburg Institute of Financial Economics Working Paper No. 02.59.

Boardman, Anthony, David Greenberg, Aidan Vining, and David Weimer. 2001. *Cost–Benefit Analysis*. Upper Saddle River, NJ: Prentice Hall.

Cairncross, Frances. 1992. *Costing the Earth*. Boston: Harvard Business School Press.

Diamond, Peter, and Jerry Hausman. 1994. Contingent Valuation: Is Some Number Better Than No Number? *Journal of Economic Perspectives* 8(4): 45–64.

Freeman, A. Myrick. 2002. *The Measurement of Environmental and Resource Values*. Washington, DC: Resources for the Future.

Friedman, Milton. 1962. *Capitalism and Freedom*. Chicago: University of Chicago Press.

Geczy, Gary, Robert Stambaugh, and David Levin. 2003. Investing in Socially Responsible Mutual Funds. Wharton School Discussion Paper.

Gramlich, Edward. 1981. *Benefit–Cost Analysis of Government Programs*. Englewood Cliffs, NJ: Prentice-Hall.

Hanemann, Michael. 1994. Valuing the Environment through Contingent Valuation. *Journal of Economic Perspectives* 8(4): 19–43.

Henderson, David. 2001. *Misguided Virtue*. London: Institute of Economic Affairs.

Holliday, Charles, Stephan Schmidheiny, and Philip Watts. 2002. *Walking the Talk*. San Francisco: Better-Koehler.

Jaffe, Adam, Steven Peterson, Paul R. Portney, and Robert Stavins. 1995. Environmental Regulation and the Competitiveness of U.S. Manufacturing. *Journal of Economic Literature* 33(1): 132–163.

Lyon, Thomas, and John Maxwell. 2004. *Corporate Environmentalism and Public Policy.* Cambridge: Cambridge University Press.

Margolis, Joshua, and James Walsh. 2001. *People and Profits?* Mahwah, NJ: Lawrence Erlbaum Associates.

McWilliams, Abagail, and Donald Siegel. 2000. Corporate Social Responsibility and Financial Performance: Correlation or Misspecification? *Strategic Management Journal* (21)5: 603–609.

McWilliams, Abagail, and Donald Siegel. 2001. Corporate Social Responsibility: A Theory of the Firm Perspective. *Academy of Management Review* 26(1): 117–127.

Mitchell, Robert, and Richard Carson. 1989. *Using Surveys to Value Public Goods.* Washington, DC: Resources for the Future.

Okun, Arthur M. 1975. *Equality and Efficiency.* Washington, DC: The Brookings Institution.

Portney, Paul R. 1994. The Contingent Valuation Debate: Why Economists Should Care. *Journal of Economic Perspectives* 8(4): 3–17.

Reinhardt, Forest. 2000. *Down to Earth.* Boston: Harvard Business School Press.

Social Investment Forum. 2003. 2003 Report on Socially Responsible Investing Trends in the United States. Washington, DC: Social Investment Forum. http://www.socialinvest.org (accessed September 1, 2004).

Stone, Bernell, John Guerard, Mustafa Gutelkin, and Greg Adams. (forthcoming). Socially Responsible Investment Screening: Strong Evidence of No Significant Cost for Actively Managed Portfolios. *Journal of Investing.*

Varian, Hal. 1999. *Intermediate Microeconomics.* New York: W.W. Norton.

Viscusi, W. Kip, and Joseph Aldy. 2003. The Value of a Statistical Life: A Critical Review of Market Estimates throughout the World. *Journal of Risk and Uncertainty* 27(1): 5–76.

Waddock, S., and S. Graves. 1997. The Corporate Social Performance–Financial Performance Link. *Strategic Management Journal* 18(4): 303–319.

World Business Council for Sustainable Development. 2000. *Corporate Social Responsibility: Making Good Business Sense.* Geneva: World Business Council for Sustainable Development.

World Commission on Environment and Development. 1987. *Our Common Future.* Oxford: Oxford University Press.

Comment on Portney

Does Corporate Social Responsibility Have to Be Unprofitable?

Dennis J. Aigner

Paul Portney has written a thoughtful and provocative chapter on the economic rationale and implications of corporate social responsibility (CSR). I found it to be extremely helpful in organizing thinking about the growing phenomenon of firms going "beyond compliance" with regard to their activities relating to environmental performance, worker health and safety, and investing in local communities.

To set the stage, Portney's working definition of CSR is "*a consistent pattern, at the very least, of private firms doing more than they are required to do under applicable laws and regulations governing the environment, worker safety and health, and investments in the communities in which they operate.*"[1] This definition doesn't imply that actions taken in the name of CSR are necessarily unprofitable, in the broadest sense of the word, but a centerpiece of the conference discussion was the idea that unless they are, there's really nothing to talk about.[2] I won't engage that idea fully except to say that unless a firm's CSR activities ultimately make good business sense, enhanced social and environmental performance is not *sustainable.*

On balance, Portney's treatment of the subject is very fair, certainly a cut above David Henderson in *Misguided Virtue* (2001), who says, "CSR rests on a mistaken view of issues and events, and its general adoption by the business world would reduce welfare and undermine the market economy." Portney instead lays out the

1. I have a small nit to pick over the use of the words "private firms." I take this to mean nongovernmental rather than the public-private distinction normally used to characterize a firm's ownership structure.

2. Portney himself has a point of view about this that is revealed as we move through the chapter.

circumstances wherein CSR can be welfare-enhancing, as well as when it can't. There are, however, a few places where Portney himself overreaches.

Referring to the empirical work to date on CSR, he relies on a book by Margolis and Walsh (2001) that reviews the literature going back to the 1970s. Portney's conclusion is that "none of the individual studies derive testable hypotheses from a theoretical model of the firm ... and few of them are very clear about the specific mechanisms through which firms' socially responsible behavior is supposed to work to their financial advantage." If writing down a mathematical model and doing comparative statics or using some other method to derive testable hypotheses is what he has in mind, which is standard operating procedure in economics, then I don't disagree. But this is an awfully narrow definition of an acceptable theoretical approach. Certainly all of the best empirical studies derive testable hypotheses from some sort of notion of the firm and how it operates, and at least acknowledge that it's not altogether clear whether good CSR performance begets good financial performance or vice versa, or that both are outcomes of good management.[3]

In discussing problems with the econometric work embodied in the extant empirical studies, Portney again seems to rely heavily, if not exclusively, on Margolis and Walsh. He says that the econometric specifications employed are "often simplistic." He refers specifically to McWilliams and Siegel (2000), who reestimated a model by Waddock and Graves (1997) and, after including an additional and clearly appropriate right-hand-side variable, found that the CSR effect became statistically insignificant. Portney then goes on to conclude: "Such findings might become common if many of the empirical studies were redone in an econometrically more rigorous way." This statement itself lacks "rigor" if indeed the generalization is based upon this one example. No doubt improvements in both data and econometric approach would improve the empirical work, but it hardly follows that such findings as turning a significant CSR effect into a statistically insignificant one would be "common" as a result.

Later in the chapter, Portney makes a related statement when referring to the empirical literature: "Unfortunately, many of these studies (though by no means all) are less rigorous than one would like, and they point in no consistent direction," referencing the direction of correlation between CSR and financial performance. Earlier, he had referred to the summary of results in Margolis and Walsh, where, on the one hand, in studies where CSR is an explanatory variable (80 of 95 studies reviewed), 53 percent point to a positive relationship,[4] while on the other, in studies where CSR is the dependent variable (19 of 95 studies reviewed), 68 percent point to a positive relationship.[5] Portney then continues, "If firms are

---

3. Some recent studies that focus on disentangling the causal effects are King and Lenox (2001), Khanna and Anton (2001), and Molloy, Erekson, and Gorman (2002).

4. Of the others, 24 percent find no relationship, 5 percent find a negative relationship, and 19 percent show "mixed results."

5. Of the others, 16 percent find no relationship, and another 16 percent find a "mixed" relationship.

looking for solid and conclusive evidence that their bottom lines will improve if they devote themselves to CSR, they will be disappointed." I actually agree with this last statement if it's based on the empirical work cited in Margolis and Walsh, even though the percentages would suggest there's something going on of a positive nature. The problem is that the Margolis and Walsh "meta analysis" contains studies going back to the 1970s, many of which are simply lousy. They're lousy because the data are lousy. They're lousy because the measurement constructs for CSR are lousy. And they're lousy because the econometric models are "simplistic," to borrow a descriptor from Portney. I would argue that only since 1996 has any really decent empirical work on this subject appeared, and it focuses on the connection between environmental performance—not CSR more broadly—and financial performance. In part, I say this because the "phenomenon" of good environmental performance, as contrasted with the occasional good performer, is relatively new, dating back to the establishment of toxics release inventory reporting. Likewise, the focus on corporate environmental reporting in recent years, culminating in the Global Reporting Initiative, has sparked a raft of activity devoted to better measurement of environmental performance.

In Aigner, Hopkins, and Johansson (2003), I have included a brief "compendium" of my own covering these recent studies, some of which also are covered in the Margolis and Walsh book.[6] Based on that assessment, I conclude that there is a positive correlation between corporate environmental and financial performance on a cross-section basis for many industry sectors, and a premium is observed when one compares stock returns over time for "good" environmental performers in these same sectors. Not surprisingly, the strength of the correlation varies across sectors, as does the observed stock returns premium attributed to good environmental performance.

Even in the face of such statistical evidence, it's not clear that firms looking for justification to devote themselves to becoming better environmental performers would be persuaded by it. Of greater impact are case studies showing how others in the particular industrial sector have achieved cost savings, created new revenue streams, or reduced operating risk. The available portfolio studies and the growing amount of attention being paid to and money flowing into so-called "socially responsible" mutual funds are likely to be persuasive at some point, too, if they're not already.

Finally, Portney really goes over the top, in my opinion, when he says that the CSR debate "*cheapens* the many quite valuable things that firms do, including providing livelihoods to people," and that firms' "profit-driven activities seem *infinitely*[7] more valuable than the things firms do under the CSR umbrella." As I said at the outset, that firms go "beyond compliance" is not a sufficient indicator

6. This article also includes an application of causal modeling to investigate the connection between the adoption of an environmental management technique and economic performance in the farm sector.

7. Emphasis added.

of behavior that is not in the firm's (owner's, shareholders') best interest in terms of profitability or risk reduction. Indeed, if CSR activities are rational in this sense, then by definition they are just as valuable as any other "profit-driven activity." I would argue that the tendency for this to be the case is much higher for investor-owned companies than for privately held companies where the owners can more easily direct company resources toward activities that are not rationalized by fundamental business principles. Even then it may be hard to tell, except by tape recording the Board of Directors' meeting or via a special CSR brain scan of the owner's thought patterns.[8]

To conclude, Portney's main contributions in his chapter are as follows:

- laying out a sensible and useful definition of CSR;
- discussing CSR as a welfare-enhancing activity (economic efficiency);
- analyzing CSR as a superior means to balance the costs and benefits of enhanced performance than could be accomplished by regulation;
- discussing CSR as a means of shifting the burden for the provision of public goods to corporations; and
- raising the question, if a firm's investments in CSR are defended on the grounds of profitability or risk reduction, what's so "socially responsible" about that?

It's regarding this last point where I have my main disagreement. Unprofitable or non-risk-reducing CSR activities simply cannot be sustained in the long run. To make good environmental performance, good worker safety and health, and targeted community investments hallmarks of corporate culture and business planning is itself the socially responsible thing to do.

Acknowledgment

I wish to thank Charles Kolstad for comments on an earlier draft of this comment.

References

Aigner, D.J., J. Hopkins, and R. Johansson. 2003. Beyond Compliance: Sustainable Business Practices and the Bottom Line. *American Journal of Agricultural Economics*. 85(5):1126–39.

Henderson, David. *Misguided Virtue*. London: Institute of Economic Affairs, 2001.

Khanna, M., and W.R.Q. Anton. 2001. Corporate Environmental Management: Regulatory and Market-Based Incentives. Unpublished working paper. University of Illinois, Champaign-Urbana.

King, A., and M. Lenox. 2001. Does It Really Pay to Be Green? An Empirical Study of Firm Environmental and Financial Performance. *Journal of Industrial Ecology* 5(1):105–16.

8. I am hard-pressed to come up with an econometric way to separate profitable CSR activities from unprofitable ones based on empirical data gathered according to Portney's definition. If I am right, and it can't be done, then the distinction has no empirical significance and should be abandoned.

Margolis, J.D., and J.P. Walsh. 2001. *People and Profits?: The Search for a Link between a Company's Social and Financial Performance.* Mahway, NJ: Lawrence Erlbaum Associates.

McWilliams, A., and D. Siegel. 2000. Corporate Social Responsibility and Financial Performance: Correlation or Misspecification? *Strategic Management Journal,* 21(5):603–09.

Molloy, L., H. Erekson, and R. Gorman. 2002. Exploring the Relationship between Environmental and Financial Performance. Unpublished working paper, Miami University, Ohio.

Waddock, S.A., and S.B. Graves. 1997. The Corporate Social Performance-Financial Performance Link. *Strategic Management Journal* 18(4):303–19.

Comment on Portney

On Portney's Complaint

Reconceptualizing Corporate Social Responsibility

Daniel C. Esty

The lack of an agreed definition for corporate social responsibility (CSR) is more than a semantic issue. The uncertainty surrounding the concept makes it hard for companies to know how to respond to calls for improved citizenship. This definitional problem is indicative, furthermore, of the absence of a sound theoretical structure within which to understand demands that corporations take action in response to certain social issues. Without a clear sense of the meaning and contours of CSR, it becomes very difficult to assess the social efficiency (i.e., possible policy gains) or the private cost-effectiveness (i.e., potential company profitability or competitiveness benefits) of CSR efforts.

Portney highlights this difficulty and offers a broad-based economic critique of CSR. I build on this analysis but focus more on "what" CSR is or might be, rather than "why" it makes or does not make sense for corporations to pursue a CSR agenda. In particular, I offer five possible "reconceptualizations" of CSR that might provide a coherent foundation for the concept.

As Portney and Reinhardt also note in this book, some commentators (e.g., Holliday, Schmidheiny, and Watts 2002) define CSR by reference to "sustainable development." But such an approach builds on shaky underpinnings. Sustainable development is not devoid of content, but it is best understood as a vision and not a concrete action–defining term. Sustainable development reminds us that environmental protection and economic progress are linked (Brundtland et al. 1987). In particular, poverty, with its associated pressures for short-term thinking, must be recognized as a source of terrible environmental degradation. Sustainable development also offers a valuable reminder that important intergenerational questions are embedded within all resource utilization issues. We thus need to think about not only short-term, but also long-term social welfare maximization. Sustainable development further emphasizes the fact that environmental policy is interconnected with choices in many other policy realms, from agriculture to transportation, energy, and trade.

Sustainable development and its sister concept "sustainability" have proven, however, to be of limited utility as a practical, on-the-ground guide for action, particularly corporate action (Esty 2001). Sustainability is a bit like "profitability." It represents an overarching goal—something on which every company must focus to optimize returns over the long term. But recognizing a need to attend to sustainability does not clarify for a corporation what specifically needs to be done. The limits of sustainable development as a driving force in the corporate world can be seen in a number of cases where companies that embrace the concept are struggling to implement it (SustainAbility 2002). Shell Oil Company, for example, shut down its Sustainability Group in 2003, concluding that the relevant issues were best dealt with by environment officials on the one hand and social and community staff on the other.

If CSR is a matter of going beyond mere compliance with the law, sustainability does not clarify where or how far to go. It does not establish a clear picture of the norms that should be pursued. Fundamentally, it is not clear whether the term is meant to be used as a policy guide or a principle for private action.

The lack of normative clarity expands as we move beyond a narrow focus on "environmental sustainability" to a broader "triple bottom line" notion of sustainability, which adds a social dimension to the economic and environmental elements of the narrower concept.[1] But as companies shift to a broader focus on corporate social responsibility, the vectors of complexity and uncertainty are multiplied.[2] Indeed, the bounds of the "social" element of the triple bottom line are potentially endless, largely undefined, and not agreed upon (Esty 2001).

As Portney spells out, in the *environmental* realm, some guideposts exist as to the norms a corporation should follow. The requirements of the law provide a robust starting point. Legal expectations in the environmental domain may be complicated, but they are relatively clear. While it is not always obvious how a corporation follows through on statutory and regulatory mandates, environmental rules are relatively straightforward, measurable, and trackable.

A second normative benchmark can be drawn from the economic realm: cost-effectiveness, the efficiency of matching the marginal costs and marginal benefits of investments in environmental protection. Environmental regulation is meant to be underpinned by this sort of "social efficiency" or social welfare maximization. As Portney points out, it may be difficult in practice to identify what is socially efficient, but the conceptual starting point is clear.

1. A "triple bottom line" focus on economics, environment, and society emerged in the late 1990s. See, e.g., http://www.sustainability.com/philosophy/triple-bottom/tbl-intro.asp (accessed September 1, 2004).

2. Compare, for instance, Schmidheiny (1992) and Holliday, Schmidheiny, and Watts (2002), reflecting the shift from a focus on "development and the environment" to a "triple bottom line" of corporate social responsibility.

Whether companies should do what is socially efficient, even if it is not cost-effective from a private cost perspective, represents a more difficult question. In other words, should responsible companies curb their emissions—internalize externalities—even if the law does not require it? From a public policy viewpoint, we might wish for this sort of self-discipline from the business community. But it is far less clear that such a posture makes sense from a company vantage point. Do we really expect companies to internalize all externalities regardless of legal requirements? Won't this simply competitively disadvantage those who take this obligation most seriously? If one assumes a functioning market in corporate takeovers, won't such a posture simply lead to a shift in corporate control, at least over time, to those with the lowest commitment to CSR, who will enjoy lower costs than their more "responsible" competitors? And how do we address the fact that what constitutes an externality often is not clear? As Portney suggests, shouldn't it be up to policymakers—not business leaders—to provide the boundaries of corporate obligations to society?

A third guidepost for CSR might be long-term expectations about *evolving* legal requirements and public expectations. A company might want to anticipate, as Reinhardt suggests, changes in relative prices. Anticipation of emerging resource scarcity, changing regulatory requirements, and evolving consumer preferences all might be rewarded. In this regard, reliance on the ability to externalize costs represents an inherently unstable basis for competitive advantage. Thus a business that cannot be profitable unless pollution harms can be emitted or natural resources obtained at below-market costs is unsustainable. The long-term trend—in every country—is to implement the "polluter pays" principle and thus internalize externalities consistent with allocative efficiency and the demands of "justice" (Esty 2004). Companies therefore will benefit from moving away from products and production practices that rely on suboptimal regulation to be economically viable. From a private profit-maximization perspective, the pace at which such shifts make sense depends on the speed at which regulation "catches up" with externalities, private consumption of collective resources, or other market failures.

The issue of what responsible companies are supposed to do becomes even more debatable as one moves beyond the environmental realm into full-fledged corporate social responsibility. The breadth of issues, potential corporate roles, and financial burdens quickly becomes very far-reaching. The 2002 World Summit on Sustainable Development in Johannesburg all but collapsed under the weight of a conception of sustainable development incorporating a wide-ranging social agenda, including poverty, AIDS, and other societal ills. When CSR-driven expectations encompass issues related to education, health, labor standards, poverty alleviation, child care, and other matters of social policy, the burden on companies becomes both unclear and unmanageable. Moreover, the need to address pollution control and natural resource management concerns tends to get lost in the mix.

In this context, the concept of CSR becomes wobbly. It either devolves into the narrow Holliday–Schmidheiny–Watts emphasis on longer-term corporate self-interest or becomes a broad vision for a better society in search of normative boundaries for the corporate role in this process. Under the narrow definition, we must ask what the social utility of CSR is. There are clearly private benefits for companies that think broadly about the payback from environmental or social investments. As the 67 case studies in *Walking the Talk* and other scholarly analyses (Porter and van der Linde 1995; Reinhardt 2000) demonstrate, environmental and social investments can build customer loyalty, enhance employee productivity, lower capital costs, reduce regulatory burdens, and improve resource productivity. But where's the social gain?

Under the broader vision, a serious question emerges about where the norms that undergird the vision are to come from. Do we really want corporations guided by something other than the law? What if companies "get it wrong" and devote their CSR energies and dollars to issues that are not public priorities? Given the diversity of the business community and the limited foundation for any assumption of an alignment between the spending priorities of corporate executives and the needs of public policy, unbounded CSR, which leaves it up to each company to determine its own sense of responsibility, will be, almost by definition, inefficient from a public policy perspective.

Portney's chapter offers a sweeping definitional, economic, and practical critique of CSR as it currently is conceived—yet concludes that CSR might be welfare-enhancing. Portney's conclusion is based on an assumption that companies generally will pursue CSR of the narrower and more self-interested kind. He warns that the more ambitious versions of CSR are likely to have negative social welfare consequences.

Perhaps a more sweeping effort to "reconceptualize" corporate social responsibility is in order. I envision five directions this reconceptualization could go. CSR could serve as (1) a check on "regulatory failure"; (2) a mechanism for regulatory flexibility; (3) a procedural guide to corporate interaction with stakeholders; (4) a way to hold corporations "accountable" for their impacts on society; or (5) a "winner's tax."

Seeing CSR as a tool for identifying regulatory failures builds on the burgeoning literature over the breadth of opportunities for win–win investments that both improve pollution outcomes and strengthen a company's profitability or competitive position.[3] Al Gore (1992) argued in *Earth in the Balance* that win–win opportunities were abundant. Others (Porter 1991; Porter and van der Linde 1995; Esty and Porter 1998) refined this logic and helped frame the so-called Porter hypothesis (Porter 1991), which suggests that in a dynamic business setting, under the right circumstances (e.g., performance-based regulation and not

3. This literature includes: Porter (1991); Gore (1992); Schmidheiny (1992); Hart (1997); Reinhardt (2000); Holliday, Schmidheiny, and Watts (2002); Doppelt (2003); and Gunningham et al. (2003).

technology mandates), environmental requirements may spur corporate innovation. Palmer, Oates, and Portney (1995) and others (Walley and Whitehead 1994) dispute this conclusion, suggesting that opportunities for "innovation offsets" are likely to be rare. They conclude that the degree of industrial inefficiency would have to be substantial for the Porter hypothesis to make sense.

One way to reconcile these divergent views is through recognition of the potential role of market failures beyond externalities, particularly information failures. To the extent that inefficiencies go unrecognized—and unregulated—a CSR focus might highlight opportunities for improved resource allocation. In particular, where market failures have not yet been addressed, CSR-driven actions may serve to cast a spotlight on unregulated pollution harms or underpriced natural resources, inviting governmental action to fully internalize the externality. CSR thus would operate as a check on regulatory failures.

Second, CSR could be seen as a way to provide firms with a degree of regulatory flexibility. In any industry, the emissions effects and costs, regulatory options, and pollution control capabilities of individual companies will vary, perhaps widely. Regulators rarely know the optimal pollution control strategy or tools for each firm or facility. Indeed, it is this sort of heterogeneity and uncertainty that provides a logic for market-based regulatory instruments such as pollution fees or tradable emissions allowances, which decentralize decisionmaking in the pollution control domain.

CSR could be understood as a similar decentralization strategy. Given the persistence and significance of information gaps (Esty 2004), it makes sense to allow firms to experiment with different control measures, so long as they meet minimum performance standards set by the government. Even under command-and-control style regulation, or perhaps especially under more rigid traditional regulatory approaches, a degree of flexibility is likely to be both socially beneficial and conducive to private cost savings through reduced regulatory burdens. Firms are likely to pursue measures that exceed government requirements when it is cost-effective for them to do so—making any such measures social welfare maximizing.

Third, CSR might be understood as laying out some procedural norms about how companies need to interact with the communities in which they operate, various elements of civil society, and the world more generally. CSR thus could be reconceptualized as a guide to stakeholder relations and a means of ensuring that companies systematically engage in a dialogue with the full range of individuals and interests touched by their activities. In this regard, CSR-driven pressures lead companies to meet certain standards of transparency and disclosure with regular reporting on their performance against environmental and social benchmarks. Centered on structuring business relationships with various stakeholder groups, corporate social responsibility might generate "engagement" that delivers real social utility, if not social efficiency.

Fourth, CSR-generated outreach and dialogue provide a way to identify and address externalities that arise from corporate impacts on the communities in which they operate or on society more broadly. As such, CSR offers a mechanism for "accountability" and "voice" for those who are affected by a corporation's actions and choices, but are not provided with opportunities for feedback through the marketplace. Because corporations, especially large multinational companies, have substantial economic throw weight and potentially vast reach, they are—or can be—powerful actors in society. CSR therefore might ensure that the implications of corporate decisionmaking, which may not be recognized or internalized by regulatory authorities, are subject to review and evaluation (Grant and Keohane 2004). CSR, under this conceptualization, serves as a discipline on the exercise of power by business leaders—part of a broader fabric of "checks and balances." To the extent that the stakeholder conversations induce companies to do more to "give back" to these communities or society at large, CSR may result in more complete internalization of presently unrecognized externalities and thus improved social efficiency.

Careful thinking about "accountability" also may highlight issues related to public expectations and environmental norms. As norms evolve and crystallize, they establish the baseline terms of a company's social license to operate (Gentry 1999). As Portney mentions, companies already do more than is required by law, such as provide health-care benefits to workers. If the business community were suddenly to walk away from this "responsibility," public outrage would ensue. CSR therefore provides a forum in which to debate and refine the scope of corporate obligations to those who are touched by the firms' activities. This dialogue must be continuous, as public expectations are dynamic. Indeed, the zone and reach of corporate obligations to society will evolve as a function of a number of factors, including economic development and rising prosperity, changing views about the proper role of the state, and public tolerance for various kinds of risks.

Finally, corporate social responsibility could be reconceptualized as a winner's tax. A number of governmental interventions, such as trade liberalization, enhance social welfare broadly but create losers within each society as well as winners. There is often talk about the need for those who benefit from policy choices to compensate those who don't. But as Stiglitz (2002) and others have argued, far too little attention has been paid to compensating the losers with the gains from the winners. So we end up with outcomes that are Kaldor–Hicks superior but not Pareto superior—with welfare advancing from a societywide utilitarian perspective but with potential negative distributional consequences as some individuals are made worse off.

In this regard, corporate social responsibility might serve as a way to press society's most economically successful enterprises to redistribute some of their winnings to those who are doing less well. As a general matter, it is the largest and most successful companies in each country or community that bear the heaviest burden of expectations on the CSR front. Thus CSR operates as a "winner's tax," imposing, in effect, a higher tax rate on a society's biggest gainers from the policy process. Whether or not this outcome is socially efficient, many would argue that it is fair and appropriate. And by addressing equity concerns,

CSR might be seen as preserving the scope for further policy actions that enhance efficiency and promote social welfare broadly but not uniformly. In the trade policy context, for example, a mechanism that takes from the winners a bit of their gains and redistributes them to the losers could serve to blunt the growing backlash against globalization and further trade liberalization (Esty 2002).

As Portney suggests, the process of bringing rigor to CSR has begun. But much more work needs to be done to make the concept a meaningful guide to action in the corporate context. Most notably, more solid theoretical foundations must be laid if we are to have confidence that CSR is a good thing from a public policy or social efficiency perspective. The five reconceptualizations of CSR introduced above offer a modest first step in this regard.

References

Brundtland, Gro Harlem, and WCED (World Commission on Environment and Development). 1987. *Our Common Future*. Oxford: Oxford University Press.

Doppelt, Bob. 2003. *Leading Change toward Sustainability: A Change-Management Guide for Business, Government, and Civil Society*. Sheffield: Greenleaf Publishing.

Esty, Daniel C. 2001. A Term's Limits. *Foreign Policy* (September–October): 74–75.

Esty, Daniel C. 2002. The World Trade Organization's Legitimacy Crisis. *World Trade Review* 1(1): 7–22.

Esty, Daniel C. 2004. Environmental Protection in the Information Age. *New York University Law Review* 79(1):115–211.

Esty, Daniel C., and Michael Porter. 1998. Industrial Ecology and Competitiveness: Strategic Implications for the Firm. *Journal of Industrial Ecology* 2(1): 35–43.

Gentry, Bradford S. 1999. *Private Capital Flows and the Environment: Lessons from Latin America*. London: Edward Elgar.

Gore, Albert, Jr. 1992. *Earth in the Balance: Ecology and the Human Spirit*. Boston: Houghton Mifflin.

Grant, Ruth W., and Robert O. Keohane. 2004. Accountability and Abuses of Power in World Politics. Unpublished manuscript presented at Yale University Department of Political Science Seminar, April 1, 2004, on file with the author.

Gunningham, Neil, Robert A. Kagan, and Dorothy Thornton. 2003. *Shades of Green: Business, Regulation, and Environment*. Stanford, CA: Stanford University Press.

Hart, Stuart L. 1997. Beyond Greening: Strategies for a Sustainable World. *Harvard Business Review* 75(1): 66–76.

Holliday, Chad, Stephan Schmidheiny, and Philip Watts. 2002. *Walking the Talk: The Business Case for Sustainable Development*. Sheffield: Greenleaf.

Palmer, Karen, Wallace E. Oates, and Paul R. Portney. 1995. Tightening Environmental Standards: The Benefit–Cost or the No-Cost Paradigm? *Journal of Economic Perspectives* 9(4): 119–132. Reprinted in *Economics of the Environment: Selected Readings*, 4th ed., edited by Robert N. Stavins. New York: Norton.

Porter, Michael. 1991. America's Green Strategy. *Scientific American* 264:168.

Porter, Michael E., and Claas van der Linde. 1995. Green and Competitive: Ending the Stalemate. *Harvard Business Review* 73(5): 120–155.

Reinhardt, Forest L. 2000. *Down to Earth: Applying Business Principles to Environmental Management*. Boston: Harvard Business School Press.

Schmidheiny, Stephan. 1992. *Changing Course: A Global Business Perspective on Development and the Environment*. Cambridge, MA: MIT Press.

Stiglitz, Joseph E. 2002. *Globalization and Its Discontents*. New York: Norton.

SustainAbility. 2002. *Trust Us. The Global Reporters 2002 Survey of Corporate Sustainability Reporting*. London: SustainAbility.

Walley, Noah, and Bradley Whitehead. 1994. It's Not Easy Being Green. *Harvard Business Review* 72(3): 46–61.

Summary of Discussion on Corporate Social Responsibility and Economics

This discussion focused on the definition of corporate social responsibility (CSR). As several participants pointed out, the question of definition is of more than semantic interest. Answers to the core questions of this book hinge on this definition. Two topics dominated the discussion. The first was the adequacy of the definition offered by Paul Portney—that CSR consists of "firms doing more than they are required to do under applicable laws and regulations," or, in the language adopted at the workshop, the beyond-compliance definition. The second major topic was the wisdom of incorporating into the definition the requirement—or at least the possibility—that corporations sacrifice profits in the pursuit of environmental goals. It was noted that the definitions under consideration may be specific to CSR in the environmental realm. Some workshop participants chafed under this constraint, particularly in light of corporate scandals, arguing that any definition of CSR should emphasize corporate governance.

Beyond-Compliance Definitions

The group largely agreed with Portney's assessment that the primary advantage of the beyond-compliance definition is that, in contrast with many other definitions in circulation, it leaves relatively little to the imagination. Cary Coglianese qualified this point by noting that compliance is not a binary variable. The law is often unclear about the status of corporate actions that harm the environment, and even when it is clear, firms may find it difficult to forecast potential tort liability and may undertake actions to limit exposure to civil lawsuits. Thus, as a practical matter, it may prove impossible to judge whether a given firm is truly beyond compliance.

Thomas Lyon defended the beyond-compliance definition against this critique, and argued that the task of the analyst is to understand why firms exceed requirements of statutes and regulations. He advocated a more careful examination of the incentives that affect firm behavior, and suggested that although avoiding tort liability may explain some beyond-compliance behavior, other incentives—such as influencing future regulation—also may prove to be important. Lyon suggested that this line of inquiry could give rise to a public-choice perspective on CSR that could help explain both why firms engage in CSR and how CSR affects environmental policy.

The group also addressed briefly the question of actual rates of compliance and overcompliance with environmental regulations by large firms in the United States. James Krier suggested that firms would do well merely to comply with environmental laws, but Portney indicated that a robust empirical literature shows quite clearly that overcompliance is the norm among large firms. Less evidence is available about the small and medium firms, which, according to Dennis Aigner, account for more than half of U.S. economic output.

Profit-Sacrificing Definitions

The major argument in favor of incorporating the requirement that firms sacrifice shareholder value into the definition of CSR is that a weaker definition—one that requires only beyond-compliance behavior—renders trivial the legal ("may they") and normative ("should they") questions. If an investment is both profitable relative to other investments and not illegal, then firms may and should do it. This view raises the question of whether firms ever sacrifice profits in the public interest, a question that several participants answered in the negative.

Einer Elhauge argued for embracing both definitions of CSR, while distinguishing between them, stressing that the best definition depends on the purpose of the inquiry. He noted that if one is an environmental activist, then the best definition might go beyond profit-sacrificing behavior, because the best way to persuade corporations to engage in such behavior would be by showing that it increases profits, whereas if one were analyzing legal or normative issues, then the definition should exclude such profit-maximizing behavior, because it raises no issue of interest. Thus he proposed that the general-purpose definition of CSR should go beyond profit-sacrificing behavior as Portney's did, but should distinguish between profit-enhancing and profit-sacrificing versions of responsible behavior. Elhauge noted that in his own essay on the legal and normative issues, he limited his analysis to profit-sacrificing versions. A number of participants were quick to point out that profit-sacrificing behavior is exceedingly difficult to observe. In particular, firms may sacrifice short-term profits in the pursuit of long-term profits. An example of such behavior cited by Bruce Hay was the $250 billion settlement between tobacco companies and the states. Although many perceived the settlement as socially responsible behavior, Hay argued that "it was quite obvious to many that [the companies were] essentially buying off the states and making them tobacco's business partners." In support of Elhauge's position, Robert Stavins pointed out that firms leave behavioral trails that economists can observe, for example in the stock market.

William Hogan offered a different vision of what it might mean for firms to sacrifice profits. He suggested that an action that appears to be profit-sacrificing in a static "prisoner's dilemma" interaction between firms might, in fact, turn out to be profit-enhancing in an iterated prisoner's dilemma game. He proposed that CSR be defined as actions that are welfare-enhancing in the aggregate and that constitute an iterated Nash equilibrium. Charles Kolstad suggested that the dynamics of such a game might convey substantial advantages to the first mover.

The group also considered briefly instances where environment and worker protection measures appear costly ex ante but are revealed to be profitable ex post. Kolstad found it perplexing that such reversals of fortune could occur repeatedly. The group reached the consensus that definitions of CSR that require such hidden cost savings are primarily of interest to advocates.

Other Issues

David Vogel and Hay both pointed out that the multiple objectives that characterize CSR could lead to situations where firms undertake investments that promote one interest at the expense of another. For example, a measure designed to increase workplace safety could swell emissions of some environmental pollutant. Hay argued that the definition of CSR should require that the beyond-compliance investments of firms be welfare-enhancing; Portney questioned whether firms should be judged on outcomes or intent.

Eric Orts suggested that CSR should be defined—at least partially—as firms conforming with deontological ethical constraints, a proposition that received considerable discussion following Forest Reinhardt's paper.

The discussion about the appropriate definition of CSR gave rise to several comments about the purpose of a definition. Three potential uses for a definition emerged, each of which implies different criteria for evaluating the definitions discussed above. For empirical research, Lyon and Coglianese emphasized that the beyond-compliance definition seems the most feasible. But Elhauge suggested that a definition that separates out profit-sacrificing behavior is more appropriate when considering normative and legal questions. Finally, a definition that asserts that beyond-compliance behavior is often profitable appears to be the de facto definition used by nearly all advocates of CSR. This view was not defended by workshop participants, but it was recognized as an important feature of the CSR landscape.

PART III

The Business Perspective

Forest L. Reinhardt

Comments

Eric W. Orts

David J. Vogel

Environmental Protection and the Social Responsibility of Firms

Perspectives from the Business Literature

Forest L. Reinhardt

Significant numbers of firms make investments in the provision or maintenance of environmental quality beyond the levels currently required by the laws of the countries in which they operate. The firms' own executives and consultants, along with business academics and other observers, have generated a large body of literature that seeks to explain the causes and effects of these discretionary investments.

Much of this writing attempts to justify these discretionary investments on normative grounds, to argue that more firms should make investments of this type, or both. Normative justifications of investments in environmental quality beyond what is required by law can be based on arguments that such investments enhance shareholder value ("it pays to be green"). Such normative justifications also can be based on ethical arguments independent of profit maximization ("it's the right thing to do"). Interestingly, much of the writing on the subject from businesspeople and business scholars makes both of these arguments at once.

In this chapter, I examine some of the writing that businesspeople and scholars at business schools have produced on the interrelated topics of corporate social responsibility, sustainable development, and the environmental performance of firms. I point out some of the findings from this literature and highlight questions that it so far has left unanswered. Finally, I sketch out the implications of the literature for each of several groups, including environmental activists, regulators, investors, executives, and business scholars.

The literature discussed in this chapter is unusually heterogeneous. It includes articles in journals with which environmental economists are familiar; these papers, in structure, method, and tone, are similar to those found in other economics journals. It includes articles in journals from the management literature, which deliberately employ the language and methodology of the social sciences. But it also includes articles from practitioner-oriented publications such as the *Harvard Business Review*, as well as books and monographs by senior corporate executives, officials of business groups, and management consultants. In these publications, formal modeling tends to be absent. Some of the authors explicitly invoke an economic model, and many more implicitly draw on economic logic, but in comparison with the economics and management literature, these practitioner-oriented pieces lean more heavily toward narrative and rhetoric, and tend to rely for their persuasiveness less on formal models and hypothesis testing than on the expert managerial judgments of the writers. Some writers in the management literature may prefer this style of argument to conventional social science methods in part because the quantitative data that might be brought to bear on the questions are of uneven quality or unavailable altogether, a problem discussed in more detail below. In addition, much of the literature on the topic of corporate social responsibility is meant to persuade someone to change his or her behavior, and narrative, rhetoric, and appeals to expert judgment are bound to play roles in such efforts. By themselves, the managerial papers and the statements by corporate executives about what they think they are doing and why might seem merely confusing. But when read in combination with other parts of the literature, they can be not only interesting, but also useful in making sense of current ideas about corporate social responsibility and their implications for firms' environmental performance.

Whether normative or positive, much of the writing on business and the environment focuses on the relationship between shareholder value and some other social objective. For example, writers may argue from logic or case-based experience that shareholder value maximization is an objective compatible with, say, sustainable development, somehow defined. Other writers may attempt to find a statistical relationship between a financial measure of firm performance and a measure of the environmental performance at the firm level. Still others may attempt to discover why managers might try to contribute to some broader social welfare measure or to see how those choices relate to other characteristics of the managers or their firms. For any of these purposes, the writer needs to decide what that broader social objective is and how, if at all, it ought to be measured. To start a chapter with definitions may seem pedestrian, but it is critical to understand what it is we are talking about, especially because "sustainable development" and "corporate social responsibility" appear to mean so many different things to different people. Hence I'll begin with a discussion of some of the ways in which "sustainable development" and "corporate social responsibility" have been defined.

Definitions

Of special interest are the definitions employed by the big firms' most famous coalition in the arenas of sustainable development and corporate social responsibility: the World Business Council for Sustainable Development (WBCSD), headquartered in Geneva. Chaired as of 2003 by Sir Philip Watts of Shell, its executive committee includes the chairmen or presidents of such firms as Dow, DuPont, Toyota, and Aventis. As of 2003, its members included 88 firms from Europe, 42 from the United States and Canada, 21 from Japan, and 5 from Australia and New Zealand; the rest of the world accounted for 16 members, of which 2 were from China and none from India.[1] The council has been prominent in debates about corporate social responsibility and sustainable development since it was founded in the run-up to the Rio summit in 1992. The senior executives of its member companies have invested considerable time in talking both among themselves and with other groups about their approaches and attitudes toward these notions.

Sustainable Development

The council's definition of sustainable development is straight from the 1987 report of the World Commission on Environment and Development, commonly called the Brundtland report after its Norwegian chairperson: "We define sustainable development as forms of progress that meet the needs of the present without compromising the ability of future generations to meet their needs."[2] Other prominent advocates of sustainability in firms invoke or explicitly cite the Brundtland definition. For example, in a prizewinning *Harvard Business Review* article, Stuart Hart (1997, *67*) wrote that "The challenge is to develop a sustainable global economy: an economy that the planet is capable of supporting indefinitely."

The Brundtland definition of sustainable development invoked by the WBCSD, although intuitively appealing, is frustratingly imprecise and difficult to test. It is, however, conceptually compatible with long-standing economic definitions of sustainability first explored in the aftermath of the first oil shock by Robert Solow (1974; 1991), Martin Weitzman (1976), and John M. Hartwick (1977). Sustainable development at the level of a nation is development that leaves the total stock of national wealth—natural plus human-made—intact. Relatively rapid drawdowns of particular natural resource stocks can be consistent with sustainability if the proceeds from the resource sales are invested in productive human-made or human capital. Conversely, even parsimonious levels of resource use can be unsustainable if accompanied by low levels of investment in human-made capital. World Bank researcher Kirk Hamilton (2000a, 2000b) has estimated per capita wealth changes for a number of countries. Jeffrey Vincent (2001) has related changes in national wealth to changes in national income, showing that countries that consume the rents from their natural resources tend

1. Membership data and much of the information quoted in this chapter can be found at the WBSCD website, www.wbcsd.org (accessed August 29, 2003).

2. Ibid.

to suffer, in the long term, compared with those that increase aggregate capital stocks over time.

Despite its elegance and intuitive appeal, this economic approach to sustainable development has attracted little attention in the management literature. The leaders of the WBCSD write that the "main message [of the concept of sustainable development] is that in thinking about environment and development issues, as in thinking about one's own life, one must figure out how to live off interest and not capital" (Holliday et al. 2002, *14*). But neither the organization nor its most prominent members have publicized any efforts to determine whether their own operations would pass tests analogous to those conducted at the national level by Hamilton and Vincent (see Reinhardt 2003). One might argue that whether the operations of a particular firm are sustainable from an economic standpoint is not an interesting question, as for a national economy to be sustainable, it is not necessary for each individual actor within the economy to maintain undiminished total assets. But the WBCSD executives do not make this argument. Given the attention that many corporate executives pay to "sustainability," and given the amount of attention they direct to their attempts to "walk the talk," it seems surprising that they do not seem more interested in conducting these tests.

Other writers on sustainability also have declined to draw on the economic literature in order to make the Brundtland idea more concrete. For example, Stuart Hart and Mark Milstein (2003), in an *Academy of Management Executive* article called "Creating Sustainable Value," note that "We use the terms 'global sustainability,' 'sustainable world,' and 'sustainable development' interchangeably to refer to global-scale drivers of sustainability. Similarly, we use the terms 'sustainable enterprise,' 'corporate sustainability,' and 'enterprise sustainability' to refer to firm-level strategies and practices to build value by moving toward a more sustainable world." But Solow, Hartwick, and Weitzman are not cited in the article's endnotes.

These economists also are absent from the 140 references in "Buried Treasure: Uncovering the Business Case for Sustainability," written by London consultants SustainAbility and copublished with the United Nations Environment Programme (UNEP) in 2001. SustainAbility (2001, *8*) quotes the Brundtland report's definition of sustainable development, then continues:

> While a definitive description of [sustainable development] may remain elusive, it is generally agreed that sustainable development involves economic, environmental and social aspects of development. SustainAbility has tended to describe a company's role in sustainable development as meeting the challenge of the 'triple bottom line'.... This concept urges companies to assess their value added (or destroyed) across environmental, social, and economic (as opposed to purely financial) dimensions.

Corporate Social Responsibility and Corporate Social Performance

"Sustainable development" as currently used is nebulous and difficult to operationalize. Dan Esty (2001) calls it "a buzzword devoid of content." Definitions of corporate social responsibility are more nebulous still. David Baron (2003) introduces his textbook chapter on the subject by noting:

> Many firms ... attempt to serve directly the needs of their stakeholders, or, more broadly, of society. For these firms, successful performance requires not only compliance with the law and public policies but also requires fulfilling broader responsibilities. Firms make charitable contributions, provide direct assistance to community organizations, support schools, provide employee and community education programs, establish programs to aid the disadvantaged, and take measures beyond those required by law to protect the environment and the safety of employees and customers.

When a writer talks of corporate social responsibility, he or she could be talking about any or all of the activities that Baron lists, and even this list is representative rather than comprehensive.

Baron (2001, 2003) draws an important distinction between corporate social responsibility (CSR) and corporate social performance (CSP). CSP involves "a redistribution from the firm to the public" (Baron 2001, *11*). CSR involves redistribution that is motivated by normative principles (Baron 2001, *8*): "both motivation and performance are required for actions to receive the CSR label" (Baron 2001, *9*). Baron calls CSP that is motivated by shareholder interest "strategic CSR": "a firm motivated only by profits many adopt a practice labeled as socially responsible because it increases the demand for its products [or, as Baron argues elsewhere, because it blunts the threat of adverse regulation]. This strategic CSR is simply a profit-maximization strategy motivated by self-interest and not by a conception of corporate social responsibility" (Baron 2001, *9*).

As Baron points out (2003, *659*), managerial motive is difficult to observe. It is not easy, particularly from outside the firm, to enumerate the costs and benefits of a particular investment, for example, in pollution control technology beyond what is required by law. Even if one could show after the fact that such an investment was destructive of shareholder value, this fact would not demonstrate that the firm was engaged in CSR as defined by Baron. Many of the benefits to shareholders of voluntary wealth transfers are deferred, contingent, intangible, or all three. So a firm that destroyed shareholder value by investing in extra pollution control equipment may not have been altruistic. It might have misread the costs and probabilities ex ante; or it might have assessed these quantities accurately and then made bets that just didn't work out. Bostonians know that Red Sox manager Johnson never should have taken out Willoughby in the seventh game of the 1975 World Series, and that Little should not have let Martinez pitch in the eighth inning of the seventh game of the 2003 New York series, but the managers' decisions do not prove that they intended to lose.

Most writers on the topic of corporate social responsibility do not observe Baron's distinction between CSP and CSR. An exception is the Worldwide Fund for Nature (WWF) (n.d.), which complains that "[t]oo much of what companies call a proactive approach to sustainable development is not really proactive. Instead it is nothing more than a part of either a risk strategy or a branding exercise." That is, WWF points out that the companies are engaging in CSP and calling it CSR. More commonly, writers use CSR exclusively, eschewing CSP. This could be because they find it advantageous to conflate the two or because they think the distinction is unimportant.

The WBCSD provides a succinct statement of its definition of corporate social responsibility: "We define CSR as 'business' commitment to contribute to sustainable economic development, working with employees, their families, the local community, and society at large to improve their quality of life.'"[3] The council thus uses the idea of sustainable development to define corporate social responsibility: CSR and sustainable development are directly linked. The council's recent book *Walking the Talk* (Holliday et al. 2002), subtitled *The Business Case for Sustainable Development,* touts itself on its cover as "The most important book on corporate responsibility yet published," again using CSR and sustainability as synonyms. At the same time, the council draws no line between corporate social responsibility and corporate social performance.

Similarly, SustainAbility, the London consultancy and collaborator with both the UNEP and the International Finance Corporation (IFC), uses the terms "sustainable development" and "corporate social responsibility" interchangeably (2002, 7): "Sustainability is sometimes known as 'corporate social responsibility' or 'corporate citizenship.' Though we use sustainability here, we accept that in many respects the terms are synonymous. They cover the same broad aspects of business: good governance, treatment of employees, impact on the environment, impact on local communities and business relationships with suppliers and customers."

Notice that nothing in these definitions of sustainable development and CSR necessitates any altruism or voluntary wealth transfers. However, in other writings, the WBCSD makes clear that voluntary wealth transfers are exactly what the members have in mind when they talk about CSR. In other words, for WBCSD, as for most people, CSR implies some beyond-compliance behavior that transfers value in the short term from the owners of the firm to some other individual or group. The council members have written:

> We hear about Corporate Social Responsibility. How do firms put it into practice? Business is not divorced from the rest of society. How companies behave affects many people, not just shareholders. A company should be a responsible member of the society in which it operates. That means contributing to sustainable development by working to improve quality of life with employees, their families, the local community and stakeholders up

3. Ibid.

> and down the supply chain. It can mean a new kindergarten, a new clinic, health insurance, playgrounds, football pitches, it can mean biodegradable packaging, cleaner fuel for trucks, sortable plastic bottles. It's a huge subject.[4]

Every one of these examples involves a voluntary wealth transfer: to local schoolchildren, to soccer players, to pedestrians and others who breathe urban air, to people who have invested in recycling systems. Every one of these examples also entails an opportunity cost: the resources that go into the football pitch or cleaner trucks could have been used for something else, including shareholder dividends. And in at least some of these examples, other groups besides the shareholders of the firm may be disadvantaged in the short term by the firm's wealth transfers. These issues are critical to the central question of this chapter: when is it in shareholders' interests for the managers of the firm to engage in such transfers?

Links to Competitive Advantage?

The members of the WBCSD say that both CSR and sustainable development are positively connected to business advantage. "We are convinced that a coherent CSR strategy, based on integrity, sound values, and a long-term approach offers clear business benefits to companies and contributes to the well-being of society."[5] They also maintain:

> Pursuing a mission of sustainable development can make our firms more competitive, more resilient to shocks, nimbler in a fast-changing world, and more likely to attract and hold customers and the best employees. It can also make them more at ease with regulators, banks, insurers and financial markets. Sustainable development policies will be profitable, but our rationale is not based solely on financial returns. Companies comprise, and are led by, and serve people with vision and values. In the long-term, companies that do not reflect these people's best vision and values in their actions will wither in the marketplace.[6]

Hence the WBCSD sees a "business case for sustainable development": "acting sustainably positively affects the bottom line of companies,"[7] and CSR makes "good business sense."[8] Or, again, "For any company, giving a high priority to CSR is no longer seen to represent an unproductive cost or resource burden, but, increasingly, as a means of enhancing reputation and credibility among stake-

4. Ibid.
5. Ibid.
6. Ibid.
7. Ibid.
8. Corporate Social Responsibility: Making Good Business Sense (Geneva: WBCSD, January 2000). Accessible through www.wbcsd.org.

holders—something on which success or even survival may depend. Understanding and taking account of society's expectations is quite simply enlightened self-interest for business in today's interdependent world."[9] Note the universality of this last claim: the council is not talking just about its own members, but about any company. The heavyweight firms on the council are echoed by the officials of the IFC, who assert categorically that "a commitment to sustainable approaches is also a wise business decision."[10]

This line of reasoning, if borne out by the evidence, reintegrates the corporate social responsibility and corporate social performance that Baron finds it necessary to separate. If it always pays to be green, then there is no actual need for corporate social responsibility. Executives can deliver corporate social performance and also maximize the private value that they deliver to shareholders. The public policy implications are enormous as well: if it pays to be green, then the use of state power to compel environmental public good provision may be unnecessary.

The outlines of the debate about the relationship between environmental quality and shareholder value have broadened considerably since a paper by Michael Porter and Claas van der Linde appeared next to one by Karen Palmer, Wallace Oates, and Paul Portney in a 1995 issue of the *Journal of Economic Perspectives*. Porter and van der Linde argued that, as a general matter, it was likely to be in firms' interests to reduce their resource consumption and pollution, as the search for gains would produce "innovation offsets" that would lower the costs to the firms. This idea has come to be known as the "Porter Hypothesis." Palmer and her coauthors disagreed, arguing that only strategic behavior involving the firms and their regulators, or the existence of implausibly massive inefficiencies in current industrial operations, could account for the benefits that Porter and van der Linde claimed. Since then, business scholars, including economists, and executives have articulated other reasons to suppose that it might pay to be green, at least some of which have been greeted with skepticism by other executives and economists.

Any story in which the voluntary provision of public goods contributes to shareholder value must invoke market failures besides environmental externalities. In a simple textbook world in which environmental externalities are the only departure from the assumptions of perfect competition, the story is not very interesting: firms are price takers and operate at zero economic profit, so they cannot voluntarily internalize any environmental costs. The stories get more interesting and more complicated only if some other imperfection—small numbers of firms, asymmetric information, other missing markets—is important. Hence in order to make the claim that "it pays to be green," one must implicitly or explicitly invoke some other market failure besides the environmental externalities.

---

9. Ibid.

10. International Finance Corporation. The Business Case for Sustainability. www2.ifc.org/test/sustainability/docs/TheBusinessCase.pdf, accessed September 11, 2003.

The next section of this chapter describes the rationales for what is variously called "corporate sustainability," "corporate social responsibility," and "beyond-compliance behavior." It relates these rationales to basic ideas about market failure and discusses the circumstances under which each might be applicable.

When Might It Pay to Be Green?

Taxonomies of the ways in which it pays to be green, or of the ways in which it might pay to be green, are numerous. At the most fundamental level, however, one might note that in order to provide benefits to shareholders, any investment, environmental or other, must increase customers' willingness to pay or reduce costs in some state of the world. Analyses of the way in which investments in environmental quality can deliver shareholder benefits may subdivide one or more of these categories, or may describe one or more intermediate effects that are potentially beneficial because they might assist the firm, under some or all circumstances, in increasing willingness to pay or lowering costs. But if they don't increase willingness to pay or lower costs in at least some scenario, the investments cannot deliver shareholder benefits. Put another way, if it is to be advantaged by engaging in short-term wealth transfers, the firm must recover that value from somewhere—either by increasing its revenues or by reducing the cost of its inputs, including purchased materials and energy, capital, and labor.

Increased Willingness to Pay

Under what circumstances might increased provision of public goods, or other short-term voluntary wealth transfers from the firm to other social groups, result in increased willingness to pay on the part of the firms' customers? Three sets of circumstances are commonly discussed. First, some firms may be able to differentiate products along environmental lines. Second, some oligopolistic firms may be able to use regulation to strategic advantage: if regulators impose disproportionate cost increases on their marginal rivals, such firms may enjoy willingness-to-pay increases that more than offset their own increases in costs. Third, going beyond this incremental thinking, sustainability advocates have argued that the environmental and social problems that impel activists' and regulators' interest in sustainable development and CSR will create discontinuous opportunities "of potentially staggering proportions" (Hart 1997, *68*).

Environmental product differentiation is conceptually similar to other kinds of product differentiation. In general, product differentiation strategies rely on the existence of some barriers to entry and mobility in the industry; otherwise, competitors could match the differentiation features and compete away the rents. A successful firm following such a strategy increases its costs but increases willingness to pay even further, and does so in ways that competitors cannot match. Brands convey information about the product's physical or intangible attributes and reinforce the barriers to entry and mobility.

Normally firms differentiate their products by offering greater private benefits to the customer: a sportier car, a faster computer. But companies also might try to differentiate their products by providing more public goods. By creating products whose environmental impacts are lighter at some stage of their lives—in manufacture, in use, or at the end of the useful life—companies may increase their customers' willingness to pay for the products. For example, a manufacturer of lumber may differentiate its product along environmental lines by asserting that the forests from which the wood comes are managed according to some set of environmental principles. In this case, the lumber may well be indistinguishable in use from lumber from biologically similar but differently managed forests. Hybrid vehicles are a good example of a product that is lighter in environmental impact during use but not necessarily during manufacture or at the end of its life. Manufacturers of recyclable bottles can claim to impose smaller burdens on the environment at the end of the bottles' lives, even if the bottles are made from virgin raw materials.

Some environmental differentiation initiatives seem to be vertical, others horizontal. A vertically differentiated product is preferred by all customers if offered at the same price as competing products; a horizontally differentiated product is designed to appeal more intensely to some customers at the expense of reduced appeal to others. Returning to the examples of the environmentally friendly lumber, the hybrid vehicle, and the recyclable bottles, it might seem as though environmental differentiation based on characteristics in use is horizontal, whereas differentiation based on production processes or end-of-life characteristics is vertical, and this seems to accord with intuition. Marketers should want to know in what kind of differentiation they are engaged, and more specifically, how the initiative affects willingness to pay in different market segments.

Environmental product differentiation has been difficult: the company essentially is asking the customer voluntarily to internalize some environmental cost, and the customer confronts the same problems of public goods and free riders as the firm. In consumer markets at the upper end of the income distribution in rich countries, some firms appear to be able to capture price premiums for environmental positioning of particular products or of the company as a whole. California outdoor wear manufacturer Patagonia, for example, commands substantial premiums that may be due, in part, to its activist environmental stance. But even here, Patagonia bundles environmental friendliness with other private goods—quality, intriguing product design, and so on—for which willingness to pay is better established. The company's founder wrote, "The most significant reason for purchasing Patagonia is quality ... environmental concerns, either in terms of the company's performance or characteristics of the product, were less important to customers" (Chouinard and Brown 1997, *124*). Other prominent examples of environmental marketing, such as those involving organic produce, also entail bundling a public good (reduced pesticide levels in the environment near fields) with a private good for which willingness to pay is well established (lower pesticide residues in food). This is not to say that households don't overcome free rider problems and contribute voluntarily to the provision of public goods. It is to say that firms have not been consistently successful in using that willingness as the basis for differentiation strategies.

It may be easier to find successful examples of environmental product differentiation in industrial markets. If an upstream firm's product can lower its customer's total cost, the product is likely to be attractive to the downstream firm. For example, a chemical intermediate that allows a downstream firm to produce its own output with lower environmental discharges, and hence lower treatment costs, should be appealing as long as the treatment cost savings exceed the price premium for the intermediate. Risk-management considerations within the purchasing firm can play either of two roles here. Risk aversion on managers' parts may lead them to leave small gains in expected value on the table if they entail an increased risk of production interruptions or quality problems. On the other hand, if industrial inputs reduce the risk to the firm of regulatory violations or tensions in the community, they may appeal to risk-averse managers even if their effect on the expected value of the firms' profits is slightly negative. (Managerial risk aversion is discussed in more detail below.)

A second set of circumstances under which improved environmental performance might lead to increased willingness to pay involves the strategic use of regulation. A firm with a technological advantage might, for example, want to see tighter regulation for itself and its competitors. DuPont's Freon Products Division, for example, seems implicitly or explicitly to have employed such a strategy in the late 1980s when it unilaterally announced its intention to stop making chlorofluorocarbons (CFCs) ahead of any regulatory requirement to do so. The announcement strengthened the hands of the regulators; they imposed a phaseout of CFCs, which enhanced DuPont's position in the markets for substitutes. The question is whether such opportunities are widespread. The CFC example appears unusual in several respects, especially the concentration of the industry and the clarity of the causality between the use of the products and the environmental insult. Although elegant models of the strategic use of regulation have been developed (see, e.g., Salop and Scheffman 1983), it remains unclear how widespread their application is in practice.

A third possibility for revenue increases through improved environmental performance is more radical. Stuart Hart, in *Harvard Business Review* (1997), writes of the "opportunities of potentially staggering proportions" that firms can grasp by "selling solutions to the world's environmental problems." For Hart, companies should strive not only to "lower material and energy consumption" and "reduce pollution burdens," but also to "develop clean products and technology, build the skills of the poor and dispossessed, foster village-based business relationships, replenish depleted resources, [and] ensure sustainable use of nature's economy." He concludes that although "public policy innovations (at both the national and international levels) and changes in individual consumption patterns will be needed to move toward sustainability, ... corporations can and should lead the way." On one level, Hart's article can be read as advocating the systematic anticipation of the changes in relative prices that will arise as some resources become scarcer. The question then is whether the institutional frameworks exist to make market prices reflect actual scarcity. Hart asserts that "the greatest threat to sustainable development today is the depletion of the world's renewable resources" such as forests, fish, and water; one could add climate stability. In each case, whatever depletion problems exist can be attributed exactly to the absence of a market price mechanism that sends signals to firms and households about scarcity. Given this absence, it is not clear why it would be in firms' interests to anticipate the scarcity unless one thinks that the scarcity will rapidly call into existence economic institutions, necessarily governmental, to bring about changes in the relative prices.

Reduced Costs

Turning from willingness to pay to cost, one can again distinguish between incremental arguments and discontinuous ones. Porter and van der Linde (1995) argue forcefully that environmentally driven cost savings are ubiquitous in industry. For a variety of reasons, firms do not minimize costs: "the world does not fit the Panglossian belief that firms always make optimal choices." In other words, imperfect markets outside the firm, coupled with agency problems inside, make environmental slack a fact of life in firms. By eliminating the slack, firms can generate additional economic value that can be delivered to shareholders in the form of cash, to external groups and future generations in the form of improved environmental quality, or, most likely, both. To be clear, Porter and van der Linde, when they discuss environmental cost savings, are talking about changes in firm activities that reduce costs while leaving willingness to pay unaffected, reducing willingness to pay by a smaller amount than the costs are reduced, or even enhancing willingness to pay. It is always possible, in the environmental or any other arena, to reduce firms' costs, but ordinarily firms are constrained from doing so, because in reducing the costs, they also would reduce customers' willingness to pay for their outputs.

Porter and van der Linde focus on the costs of purchased materials and energy costs. With respect to these cost categories, the debate unfortunately has not progressed very far since 1995. Advocates of the cost-savings approach cite examples of firms that have saved money through environmental initiatives. Skeptics argue that the stories are not generalizable, that one-shot gains from environmental scrutiny will not translate into continuous improvement, and that a fuller accounting that recognizes the costs of managerial time and effort might not show such a benign picture. We know from Porter and van der Linde, and numerous followers, that one can find examples of individual firms that saved money; we know from Palmer et al. that aggregate environmental spending, net of Porter's offsets, is positive. We know far less about the circumstances under which a firm is more or less likely to find environmental cost-savings opportunities, or about the ways in which returns to environmental cost-savings initiatives compare with those of cost-cutting programs of a more old-fashioned sort as effort increases.

Meanwhile, since 1995, the debate about cost savings has broadened. Researchers have raised the possibility that voluntary environmental improvements might save not just raw materials and energy costs, but also labor costs, capital costs, and the costs of future environmental compliance.

Consider labor costs. Firms with better environmental records may be able to attract better workers for lower wages. Such firms may be able to attract and retain superior workers who will deliver higher productivity either because of their innate superiority or because they will work harder for a firm whose senior executives espouse values congruent with their own. According to Lise Kingo of the Danish firm Novo Nordisk (quoted in Holliday et al. 2002, *110*): "A commitment to corporate social responsibility helps us attract and retain the most talented people. We know it particularly means a lot for young people to work for a company they can feel proud of and a company that reflects their own personal values." Kingo's assertions seem intuitively plausible, but systematic study of this prospect so far is sparse.

Similarly, with respect to capital costs, one might argue that capital providers might prefer to fund companies that provide public goods or transfer some of their wealth voluntarily, either because the capital providers are themselves altruistic, or because the capital providers think that companies of this sort offer better investments from a narrow cash-flow point of view. Investment houses have arisen to meet the demands of "social investors" who want, for one reason or another, to own equities of socially responsible companies, defined variously. Systematic study of this question is widespread, and considerable evidence exists on the relationships between social performance and capital market performance. This evidence should be of interest to investors; unfortunately the evidence is equivocal, as I will discuss below. From a managerial point of view, the question seems to be whether enough "social investors" exist to lower, on the margin, the cost of capital for the firms that such investors prefer. This seems intuitively implausible but appears to have attracted little rigorous attention.

It also might be the case that firms can reduce the expected value of their environmental costs by engaging in voluntary initiatives that preempt governmental regulation. John Maxwell, Thomas Lyon, and Steven Hackett (2000) show analytically that such preemptive voluntarism can be efficacious from the firm's point of view. Under some circumstances, it also can be welfare-enhancing, as the firms and the environmentalists need not spend resources on lobbying efforts to influence the level of regulation. Note that this story differs from the one told earlier about DuPont and CFCs, in which the firm initiated voluntary action to increase the probability of regulation. It also differs from a story in which the firm, by engaging in voluntary actions, signals to activists that it is politically weak and thereby increases the amount of company-specific activist pressure.

It remains to consider a more radical view of cost savings within the firm. Amory Lovins, Hunter Lovins, and Paul Hawken wrote in *Harvard Business Review* in 1999 that "natural capitalism," involving dramatic increases in the productivity of natural resources, a shift to closed-loop production systems, a move from selling goods to selling services, and reinvestment in natural capital, will "subsume traditional industrialism," bankrupting the firms that ignore it. For example, firms "must restore, sustain, and expand the planet's ecosystems so that they can produce their vital services and biological resources even more abundantly," and they can "make natural resources—energy, minerals, water, forests—stretch 5, 10, even 100 times further than they do today." Although they anticipate, like Hart, that prices of natural resources will rise, Lovins, Lovins, and Hawken assert that many of the innovations they propose will pay for themselves even at today's prices. Here again the question is why, given the absence of mechanisms to price adequately the natural capital that Lovins, Lovins, and Hawken discuss, firms ought to invest in increasing the productivity of those resources. The fundamental problem of the divergence of private and social cost is still with us.

Improved Risk Management

A third set of ways in which managers might create shareholder benefit through environmental protection relates to risk management. The benefits of environmental investments are contingent: the investments pay more than the opportunity cost of the capital in some future states of the world but not in others. In this respect, they are no different from other investments that firms make in product development, advertising, an improvement in the energy efficiency of a production process, or anything else: the new dog biscuit may or may not win over the consumer; the energy efficiency improvement might pay for itself if oil costs $30 but not if it costs $20. The uncertainties surrounding environmental investments may or may not be greater than those surrounding other kinds of investments; the question does not seem to have received much empirical attention.

The contingent nature of the payoffs to environmental investments does, however, raise questions about the normatively appropriate course of action when a possible investment lowers the firm's expected value but also lowers the variance around that expected value. Why should shareholders, who can diversify away the risk of new pollution control laws, unfavorable changes in consumer demand, or other adverse environmental shocks to the firm, want to pay managers to reduce this risk if the risk-reduction efforts reduce the firm's expected value? Froot et al. (1994) provide an overview. First, note that the firm's expected value also is affected by the expected costs of financial distress, which can be reduced by smoothing year-to-year cash flows; when this benefit is taken into account, some risk-management policies may pay for themselves in expected value terms. Tax considerations also may make smoother cash flows more valuable. Beyond this possibility, imperfections in the capital markets, and the absence of free flows of information about firms and their prospects, provide some possible answers. If the firm's managers know more about its prospects than do outside shareholders, internally generated capital may be cheaper than outside capital, and a firm contemplating a multiyear investment project may wish to reduce the variance in annual cash flows. Also, if managers cannot hedge their own risks of unemployment should catastrophe strike the firm, they may demand higher compensation if the firm's shareholders disallow them from insuring against it contractually. In practice, executives routinely reduce risk while reducing expected value, and shareholders are happy to let them do this. Most prominently, many large firms buy liability insurance, a purchase readily observable by shareholders. Besides the risk shifting that purchases of insurance contracts entails, managers can reduce their firm's environmental risks by making physical investments that reduce the probability of an adverse event or reduce its cost if it does occur.

Risk-management considerations are paramount in one of the most widely invoked reasons for firms to invest in public good provision: the maintenance of the social license to operate (see, e.g., SustainAbility and UNEP 2001, passim). The idea is that societies permit firms to operate subject to the firms' adherence to written rules and also to informal norms that the firm may discover only by breaching them. Societies, through their regulatory and legal apparatus, through boycotts, or even through decentralized, nongovernmental violence, can force firms to cease or radically reconfigure their operations if those operations violate the written laws or the unwritten norms. From the firm's point of view, voluntary investments in public good provision can serve as an insurance policy against involuntary cessation or changes in operations. The "license to operate" argument for CSR is at bottom an argument about risk management.

As Gunningham et al. (2003, *35–38*) point out, managers actually need to have not one but three licenses to operate. One comes from the legal authorities, who literally can revoke the license by withholding legal permission to operate. The second comes from society more broadly; this could be revoked in a number of different ways, although commonly it seems the legal authorities would serve as the agents of the society. The third license to operate comes from economic actors. For senior executives, these licensors are primarily shareholders, who can revoke the license to operate by firing the executives and hiring others to manage the assets. For junior managers, the licensors are their hierarchical superiors. If one is to explain managerial behavior, it is necessary to take into account not only the risk of revocation of the legal or social license, but also the risk of revocation of the economic one; managers surely take this last risk very seriously. And if one is to give sensible normative advice to managers, one needs to be cognizant of all three licenses. Otherwise one might conclude that because the consequences of a revocation of the legal or social license are so large, it makes sense to spend almost any amount of money to insure against it. In other words, the need to retain the economic license limits the amount of expected shareholder value that managers can destroy in the name of the protection of the social license.

It is certainly possible to engage in risk management or in contingent cost control that seems ex post to be an overinvestment. In the pulp industry, Gunningham et al. (2003, *60–74*) cite "the experience of firms that suffered economically from overestimating the stringency of anticipated regulations" (*71*). In particular, some firms anticipated that American regulatory requirements to move from "standard" pulp bleached with elemental chlorine were merely a precursor to requirements to eliminate the cheapest substitute, chlorine dioxide, and thus move from "elemental chlorine free (ECF)" to "totally chlorine free (TCF)" pulp. When the regulatory momentum stalled, firms that anticipated tighter regulations were left with unrecoverable costs.

Attempts by firms to enhance their control over decisionmaking processes that affect spending levels also can be seen as a type of risk management. Through voluntary initiatives, firms may be able to reestablish some power over decisionmaking that otherwise might be lost to other social groups. Analyzing environmental voluntary agreements (EVAs) such as EPA's Project XL and its 33/50 program, Maxwell and Lyon (2001, *347*) write that "it appears that the EPA and affected members of regulated industries may be attempting to use EVAs as a way to win back some of the power they have lost under the traditional regulatory structure. It is clear that EVAs grant these two groups much more power than the Congress or national environmental groups, both of whom are blocked from participation in many EVAs." In other words, voluntary public good provision can increase the firms' political power and hence managers' flexibility to respond to future environmental pressures in such a way as to increase willingness to pay or lower costs.

To summarize, the benefits to shareholders from beyond-compliance environmental investments by firms can arise from enhanced expected willingness to pay, from reduced expected cost, from improved risk management, or from some intermediate effect that improves one of these three things. Any taxonomy can be reduced to these four elements.

For example, environmental consultants SustainAbility, in collaboration with staff at the IFC (2002, *5*), cite six different "Business success factors" that can be enhanced by an emphasis on "sustainability": "Revenue growth and market access, Cost savings and productivity, Access to capital, Risk management and license to operate, Human capital, and Brand value and reputation." The first of these clearly relates to willingness to pay, the second to cost. The third might relate to cost, if sustainable firms can get capital more cheaply, or to the license to operate and hence to risk management, if nonsustainable firms are for some reason unable to obtain capital. The fourth is explicitly about risk management. Human capital could directly relate to cost, if sustainable firms can get workers for lower wages or with higher productivity, or to intermediate factors that lead eventually to business advantage—if, for example, sustainable firms attract better workers, who then develop cleverer marketing schemes or lower-cost manufacturing systems. Brand value and reputation are also intermediate factors that could lead to higher revenues, lower expected costs, cheaper risk management, or all three.

In their study of environmental behavior in the pulp and paper industry, Gunningham et al. (2003, *23–24*) specify "four types of beyond-compliance environmental protection measures by business firms: 1. Win–win measures ... that will increase corporate profits.... 2. Margin of safety measures that overcomply with current regulations, much as a motorist might drive 5 mph below the speed limit.... 3. Anticipatory compliance measures that overcomply with current regulations because a firm anticipates specific increases in stringency.... 4. Good citizenship measures ... justified on the grounds that enhancing the firm's reputation for good environmental citizenship will in the short or long run be good for business." Of these, the "win–win measures" must increase willingness to pay or lower costs; the examples cited by the authors are cost-focused, unsurprisingly in a business where competition on any nonprice basis is extremely difficult. The margin of safety measures and anticipatory compliance measures could reduce the expected value of long-term compliance costs plus fines, and could thus pay for themselves in expected-value terms; or they could constitute insurance that reduces expected value slightly while protecting against low-probability but expensive events. In particular, the margin of safety measures may be necessary because pollution control equipment, like other capital, cannot operate with complete consistency. And anticipating future regulatory requirements may reduce overall costs by taking advantage of economies of scale or scope in pollution control, if the government does tighten the regulations. A "good citizenship measure" can be "good for business" only by raising willingness to pay or lowering costs, at least in a contingent manner; managers may think, for example, that such a reputation will enable them to retain access to product markets or attract and retain higher-quality workers.

Taxonomies of the ways in which it pays to be green have become increasingly complex. For example, SustainAbility and the IFC (2002, *10*) devised a 7 (6 matrix designed to show all of the different business benefits of "sustainability." The rows are "Business success factors," as enumerated above: "Revenue growth and market access, Cost savings and productivity, Access to capital, Risk management and license to operate, Human capital, and Brand value and reputation." The columns list "Sustainability Factors," including such things as "Stakeholder engagement," "Environmental process improvement," and "Human resource management," by which they mean not just the traditional human resource function, but "Sound employment practices such as fair wages, a clean and safe work environment, training opportunities, and health and education benefits" (2002, *12*). The cells are hypothesized causal linkages between the sustainability factors and the business success factors. The analysts emphasize seven such links as particularly important: environmental process improvement enhances revenue growth and market access; environmental process improvement enhances cost savings and productivity; environmental improvement enhances brand value and reputation; stakeholder engagement enhances risk management and license to operate; local economic growth helps revenue growth and market access; human resource management helps cost savings and productivity; and human resource management helps human capital. At the same time, the analysts assert that 29 other links between sustainability factors and business success factors are also important; for SustainAbility and the IFC, only 6 of the 42 potential causal links are negligible.

Evidence

We have seen that business scholars have tried to map out the mechanisms by which sustainable development or corporate social responsibility in the environmental arena might deliver positive returns to shareholders, and to enumerate the ways in which it might pay to be green. Meanwhile, researchers have been attempting to learn whether these mechanisms are actually significant from an empirical standpoint. This research has used a variety of methodologies, the most important of which are statistical techniques of varying degrees of sophistication and case-based or historical approaches. I consider these in turn.

Statistical studies of environmental performance at the firm level fall into two broad categories. The studies in one group try to relate financial performance to environmental performance or social performance. That is, they use a measure of financial performance as a dependent variable and seek to explain its behavior using measures of environmental performance, commonly, but not always, including other explanatory variables as well. The studies in the second group use a measure of environmental or social performance as the dependent variable and try to explain this performance by examining other characteristics of the firm, often including some measure of financial performance. The second group thus focuses on the determinants of firms' environmental choices, and the first on the financial consequences of those choices.

Statistical Studies of the Links between Environmental and Financial Performance

The studies in this statistical literature vary considerably in quality. A variety of inescapable conceptual and practical difficulties make even the best more equivocal than one would like.

First, there are obvious problems in measuring the phenomena of interest. Measuring the financial performance of firms is far from straightforward. Most studies use an accounting measure such as return on book equity or return on book assets, but the denominators of these ratios are historic and may bear no relation to market values. Some other studies use other measures that are less useful from a financial perspective, such as return on sales. To avoid the arbitrary nature of accounting data, some other researchers have used market-based measures such as Tobin's q or cumulative abnormal returns. But these measures also can be seen as problematic, particularly since the collapse of the equity bubble of the late 1990s.

Measuring environmental performance or social performance is even more problematic. The use of output measures such as the emissions tracked by the EPA in the Toxics Release Inventory (TRI) is common; researchers can choose between levels (normalized for revenues or some other measure of size) and changes in the levels (normalized) over time. Researchers also could conceivably use input measures, but data on firms' expenditures on environmental quality can be inconsistent across firms and over time. Some researchers use measures that are thought to be correlated to overall environmental effort: results of content analysis of annual reports, for example. Often researchers outsource the measurement of environmental performance by using ratings compiled by environmental investment advisers or activists. These typically take the form of rankings on a simple (e.g., five-point) scale. An additional possibility is to take self-reported indicators of output, input, or managerial attitude. The environmental variables in the study thus can be as simple as dummies that equal one if the firm has participated in a voluntary cleanup program or claims to have instituted particular internal procedures. A final commonly applied approach is to conduct event studies around an environmental accolade, such as the reception of an award for environmentally friendly performance, or an environmental disaster, such as an oil spill.

A few minutes of reflection will suggest that any of these measures is potentially flawed. Emissions:sales ratios will vary for technological reasons across firms, even within a relatively narrowly defined industry category such as petroleum refining. Changes in emissions:sales ratios necessarily focus on a particular period, ignoring changes before and after, and possibly presenting a misleading picture of long-term environmental performance. Difficulties in obtaining measures of environmental effort that are consistent across firms and across time are all but insuperable. Ratings of investment advisers or activists may take into account outputs, inputs, or both, but the constructors of these ratings face the same problems as the academics. Using content analysis of annual reports, or responses to questionnaires about environmental effort or performance, risks conflating environmental performance with environmental rhetoric.

Data are most commonly available at the firm level. Such data aggregate the effects of the several mechanisms by which public good provision might be supposed to affect financial performance. It would be desirable, in many cases, to be able to isolate one of these mechanisms; this would require, for example, using the product introduction as the unit of analysis for a study of product differentiation, or using the cost-cutting initiative as the unit of analysis for an examination of the Porter hypothesis. But such data are not generally available.

Sampling bias constitutes a second problem. Joshua Daniel Margolis and James Patrick Walsh, after examining 95 studies published between 1972 and 2000, wrote that "over half of the 95 studies examine exemplary, notorious, or very large firms" (2001, 7). For example, some studies examine only Fortune 500 companies, others only firms in industries that are thought to be especially pollution-intensive, such as steel, oil, and pulp and paper. Questions remain about the degree to which results can be applied to a broader sample of firms. A related point is that the vast bulk of the statistical studies have been American, in large part because, although American data on environmental performance are weak, they are stronger than those available for firms in other countries.

A third problem with the statistical literature on financial performance and CSP relates to causality. Supposing that one finds a plausible measure of financial performance and a plausible measure of social performance, and that some correlation exists, questions of causality are still difficult to disentangle. Many contributions to the literature report correlations but do not even attempt to take into account other variables that plausibly affect CSP, financial performance, or both. Obviously these are not very convincing. Other studies attempt to correct for industry effects, scale, and other plausible drivers of environmental or financial performance, usually by using multivariate regression. Aside from omitted variable bias, significant questions remain about the direction of causality in any observed positive relationship between social and financial performance. Do socially responsible firms perform better because they are socially responsible, or does strong financial performance create opportunities to engage in wealth transfers? The use of lagged variables in a panel setup should, in theory, be sufficient to tease out the causal linkages between financial and environmental performance. Unfortunately, the time periods covered by the studies are often too short for the hypothesized effects plausibly to take place.

Finally, researchers need to confront the possibility of diminishing and eventually negative returns to environmental investment, and many of the studies in the literature cannot do so. One might suppose, following Porter, that firms can find some opportunities to reduce pollution profitably, or, more broadly, that some voluntary wealth transfers give rise to increased shareholder value. It would not follow that a doubling of environmental or social effort would yield twice the benefits; at some point, one would expect the marginal returns to be negative. It is hard to measure environmental effort or output very precisely, however, and many studies use insufficiently fine measures to take this possibility into account.

To see how these general problems make the statistical work difficult, consider a couple of the better examples of work of this sort. With the firm as the unit of observation, Dowell, Hart, and Yeung (2000) regressed a measure of stock market performance (Tobin's *q*) on a set of variables that included, besides R&D intensity, advertising intensity, size, and other control variables, a measure of environmental performance. Specifically, Dowell et al. constructed dummies using multinational firms' voluntary reports to the Investor Responsibility Research Center, a research group interested in companies' social and environmental performance. The dummies indicated whether the firm said that it applied a uniform environmental standard, tighter than any national standard, to all its operations; applied U.S. standards everywhere; or allowed operations in different countries to adhere instead to less stringent individual-country standards. The sign on the dummy for uniform tight global standards was positive, indicating that within the sample of firms, more aggressive environmental performance was not incompatible with superior stock price performance. Curiously, the authors found no financial performance difference between firms that applied the U.S. standard everywhere and those that applied local standards.

Dowell et al. point out that in an analysis of panel data, "previous years' environmental standards are not significant predictors of current Tobin's *q* values." Correlation is not causality. Note also that Dowell et al. did not find diminishing returns to environmental investment; in fact, their results are consistent with a minimum scale effect, but because the independent variable of interest could take only three values—local, U.S., or tight global—it is hard to know how much weight to attach to this result.

In another carefully produced study, Andrew King and Michael Lenox (2001) analyzed a panel data set for 652 American manufacturing firms during the period 1987 to 1996. They regressed Tobin's *q* on a number of variables that included emissions, as reported in the EPA's TRI. Like Dowell et al., they found positive correlations between environmental performance and financial performance, but also could not establish the direction of any possible causal links.

King and Lenox emphasize that the question of causality is less important for investors than it is for managers. Investment advisers such as Innovest in New York or SAM in Zurich postulate that aggressive environmental management and public good provision are proxies for management acumen and foresight more generally, and recommend portfolios based on this idea. King and Lenox write, "For their purposes, it matters little whether environmental performance leads to financial performance or simply provides an indicator of firms that have high financial performance. From the perspective of corporate managers and policy analysts, however, the distinction is critical .... To fully demonstrate that it pays to be green, research must demonstrate that environmental improvements produce financial gain."

King and Lenox's conclusions could serve as a summary of the literature as a whole: "Much of the variance in our study is attributed to firm-level differences. Better understanding of these differences might provide a richer understanding of profitable environmental improvement. It may be that it pays to reduce pollution by certain means and not others. Alternatively, it may be that only firms with certain attributes can profitably reduce their pollution.... 'When does it pay to be green?' may be a more important question than 'Does it pay to be green?'"

Statistical Studies of the Links between Firm Characteristics and Environmental Policy

As one way of trying to answer this question, one might ask what kinds of firms have invested in beyond-compliance pollution control. If we assume that managers understand the private costs and benefits of voluntary public good provision, and that they are interested in maximizing profits, then their behavior becomes informative. For example, Seema Arora and Timothy Cason (1995) analyzed firms' decisions about whether to participate in EPA's 33/50 program. They found that firms that are large or that have substantial environmental releases are more likely to participate. (Arora and Cason also looked for a relationship between the competitive conditions faced by the firm and its discretionary public good provision. They found that firms in unconcentrated industries are more likely to participate. This seems counterintuitive, but the Herfindahl indices were computed at the two-digit SIC level and only seven two-digit industries were represented in the sample. Obviously two firms in the same two-digit SIC code could face very different industry economics.)

Along similar lines, Irene Henriques and Perry Sadorsky (1996) sent a survey to firms in Canada to gain understanding of the circumstances that affect the probability that a company will have a formal plan or organizational structures for addressing environmental problems. More regulatory pressure, more customer pressure, and more shareholder pressure all increase the likelihood that a firm will have a plan or structure. A higher sales-to-assets ratio tends to decrease it. Henriques and Sadorsky say that this is because firms with higher asset turns are operating close to capacity, but it might be that firms with fewer turns are just in more asset-intensive businesses that require different kinds of risk management and long-term planning. Both Arora and Cason and Henriques and Sadorsky provide evidence that the fundamental economic conditions confronting the firms' managers influence environmental choices. Because these conditions obviously differ across firms, any answer to the "pays to be green" question has to be contingent on firm-specific factors.

Histories and Case Studies

For this reason, it might seem promising to use historical case studies to shed light on the contingencies. The inconclusiveness of the statistical work so far suggests that in-depth, case-specific analysis of the relationships would be useful. In particular, one might use historical case studies to shed light on either or both of the relationships that the statistical work has explored: between the circumstances confronting the firm and the firm's choice of environmental policy, and between the environmental policy and financial performance.

I myself have used case studies in the past to examine some ideas about necessary and sufficient conditions for a positive relationship between some beyond-compliance behavior and shareholder value (Reinhardt 2000). For example, I use basic ideas from marketing and strategy to try to explain why some environmental product differentiation initiatives have succeeded but others have failed. Using similar logic, I compare the environmental risk-management policies of two different pulp mills in light of their technological constraints, their cost positions, the characteristics of the nonmarket environment that they confronted, and so on. Most of my cases are American.

Work of this sort obviously cannot prove that a particular relationship exists. But the factors that influence the relationships between environmental and financial performance are likely to be firm-specific and may be difficult or impossible to capture in publicly available statistics. Under these circumstances, historical analysis, even if never completely persuasive to social scientists, can advance our understanding of the relationships.

My case analysis, rooted in strategy and economics, also could be criticized for paying insufficient attention to organizational dynamics within the firm and the cognitive biases of individual managers. For example, Cary Coglianese and Jennifer Nash (2001) argue that what they call "management commitment" is an important determinant of firm environmental performance. It is these organizational and personal factors that Andrew Hoffman emphasizes in his 1997 book, *From Heresy to Dogma: An Institutional History of Corporate Environmentalism.* Whereas my book emphasizes the relationships between external market circumstances and the competitive positioning of the firm on environmental choices and outcomes, Hoffman's is more concerned with the effects of internal managerial attitudes on those environmental choices. In particular, he traces the change in managerial attitudes toward the environment that occurred between the 1960s and the 1990s in the American chemical and oil and gas industries. When he uses the word "institutions," Hoffman, a student of organizational behavior, means "the coercive rules, normative standards, and cognitive values" of the firm's social environment (*7–8, 37*), a definition broader than the one familiar to economists. Using this broad definition, he attempts to relate the changing external pressures on the industry during this period to the changes in the organizational structures and processes of firms, and even to the cognitive processes of individual managers.

Hoffman's study is a useful step in trying to integrate economics and law on one hand and organizational theory on the other. Unlike the statistical work or the historical case analysis just cited, however, his book places little emphasis on interfirm variance in technologies, costs, and culture that might plausibly account for disparities in environmental policymaking. Indeed, he regards these disparities as relatively unimportant: "as this history reveals, industries have made organizational shifts in relative unison ... firms have behaved in a clannish fashion, evolving through waves of commonly accepted organizational trends." Hence for Hoffman, the history of oil and gas or of chemicals is "inconsistent with purely rational economic explanations, which should see different industries, as well as different firms within those industries, reaching autonomous organizational decisions based on their own cost structures and strategic objectives" (*145–46*). One might argue that this is just what one does see: the big oil companies are now adopting quite different attitudes toward environmental issues, and the chemical companies, too, are not uniform in their approaches. Thus just as work rooted in strategy and economics can legitimately be criticized for insufficient attention to the organizational, so the organizational work can be criticized for taking insufficient account of strategy and economics.

More recent work seeks to integrate the economic and the organizational. Jennifer Howard, Jennifer Nash, and John Ehrenfeld (2000) analyze American chemical companies' implementation of the Responsible Care program, a voluntary exercise in self-regulation initiated in the late 1980s. Under the program, the industry's trade association established guiding principles and management "codes." For example, under the codes on community awareness and emergency response, the first to be implemented, companies are to provide information on safety and environmental questions to residents of towns near their plants, spend time and money on emergency preparedness, and so on. But as Howard et al. point out, the companies are given considerable discretion: "the codes outline what a company must do, but not how it should do it" (*65*). Under these circumstances, both economic positioning and managerial attitudes appear to affect the ways that Responsible Care is implemented and its impact on managerial practice. While the implementation of the codes that relate to "organizational image" and community relations have been implemented relatively consistently across firms, the firms vary substantially in the implementation of codes such as process safety and distribution, where the firm's operations are opaque to the public. Managers whose firms had already taken environmental matters seriously before Responsible Care tended to take the new initiative more seriously as well: "companies that have a firm prior commitment to a certain set of practices will be more likely to conform to pressures to extend them" (*76*).

Along similar lines, Neil Gunningham, Robert Kagan, and Dorothy Thornton (2003) recently completed an intensive study of 14 pulp mills in the United States, Canada, Australia, and New Zealand that shows clearly how the interactions between external economic pressure and internal managerial attitudes shape firms' environmental choices. In their view, "The influence of external pressures on environmental performance depends on an "intervening variable"—*managerial attitude*, or more broadly, what we call *environmental management style"* (*95*). They emphasize that the license to operate is really tripartite: regulatory, economic, and social actors all hold licenses that they can revoke (see above). Any of the three license holders can act as a brake on either of the others, although it seems that the most common situation would be for the social or regulatory actors to serve as a brake on the economic, and vice versa. The relative importance of the pressures for financial performance and for social or environmental performance will differ across firms, based on the firms' corporate governance structures, past financial and environmental records, and so on. Further, different firms may assimilate and integrate these competing pressures in different ways, depending on their environmental management style.

Through extensive interviews, Gunningham et al. sought to understand managerial attitudes toward the environment. Based on these interviews, they define four categories of environmental management style. The first they call "true believers": managers who are "morally driven and almost evangelical in their pursuit of environmental excellence." The second category is "environmental strategists": these "made strategic use of corporate environmental policy and most (but not all) of them believed that the current sociopolitical climate required them to be excellent environmental performers." "Committed compliers" were similar to strategists but "tended to define all strategy and demands in reference to regulatory requirements." "Reluctant compliers" tend to "comply because they do not want to get caught" (*128–29*).

With this taxonomy in hand, Gunningham et al. relate environmental performance, as measured by pollution loadings, to management style, finding that "environmental management style was a much more powerful predictor of mill-level environmental performance than regulatory regime or corporate size or earnings" (*130*). But at the same time, "environmental management style operates within important economic constraints. It is far from omnipotent in shaping environmental performance. The correlation between environmental management style and environmental performance, while quite significant, is not overwhelming powerful ... no mill in our sample, including the True Believers, has adopted environmental improvements that were vastly better than those employed by Reluctant Compliers" (*133*).

Gunningham et al. thus have made a significant step forward in understanding how external pressures and internal managerial attitudes interact to shape firms' environmental performance in the pulp industry. Their work has the additional advantage of cross-nationality. Although it is subject, as are other case-based analyses, to questions about the degree to which one can use it as the base for broader generalizations, it represents the integration of the economic and the organizational that appears necessary to build a complete picture of environmental decisionmaking in industries for which market imperfections besides environmental externalities remain important.

Conclusions

The business literature on the environment conceivably could be of use to each of several audiences, including government regulators, environmental activists, investors, business executives, and business scholars. In this concluding section, I summarize what I see as the main lessons of this literature for each of these groups, and then sketch out some implications for future research.

Lessons for Activists and Regulators

Perhaps surprisingly, the literature may be most useful to activists and regulators who seek to move corporate behavior in the direction of greater "responsibility" or a more pronounced emphasis on "sustainability," by which these groups tend to mean a greater degree of public good provision or more extensive voluntary transfers of wealth to groups who are not the companies' shareholders. If they seek to alter managerial behavior, they need to understand it, and the literature collectively sheds considerable light on this behavior.

Managerial behavior in this arena is quite complex. Managers' motivations are not always straightforward, and the constraints that they confront, both externally and internally, are extremely complicated. Even the managers themselves do not always appear to be very clear why they are pursuing a particular course of action (i.e., what exactly it is that they are maximizing) or what actions by other parties such as government regulators or activists would serve their interests.

Some of the complexity arises because firms make investments that affect the marginal costs of later decisions, in ways familiar to students of industrial organization. Because of differences in scale, scope, and other aspects of competitive positioning, different firms in the same industry will face different constraints and therefore optimally will respond differently to any particular environmental initiative, even if they all are single-minded maximizers of shareholder value. On top of this, because of differences in governance structures, different firms in the same industry may have divergent objectives as well; a privately owned firm, for example, may have different objectives than one whose shares are publicly traded. The competitive positions of the firms within the industry themselves also can give rise to divergences in objectives. Firms with stronger market positions and larger, more stable rent streams may be more interested in voluntary wealth transfers as insurance policies against boycotts or regulation than are weaker counterparts with tighter cash constraints. In addition, organizational work suggests that divergent management styles or corporate cultures also may lead to differences in environmental policies.

The implication here is that a single "industry" position on any particular environmental topic is unlikely to arise, and this has profound implications for the strategies of regulators or activists who want to change corporate behavior. It is shareholders and managers, not industries, who have interests, and successful activists and regulators sensibly will exploit this fact by identifying and emphasizing issues for which the myth of a monolithic industry interest is least likely to be tenable. They will attempt to recruit big, rich, risk-averse firms as de facto allies. They will hope that these firms will engage in the kind of voluntary wealth transfers that please the activists and regulators. Even more, they will hope that these firms, by asserting that "it pays to be green," will intensify the regulatory and social pressure on other firms for which the wealth transfers may not be so advantageous.

The environmental path chosen by a particular firm is affected by external economic conditions, by past investments in environmental performance and reputation, and by managerial attitudes. Activists and regulators should be especially interested in research that deepens their understanding of the way these influences interact to shape corporate performance in particular industries and particular firms. Case studies of particular firms or industries that consider the evolution of corporate social performance in some detail are especially likely to be valuable.

More generally, however, this literature suggests that regulators and activists ought to be skeptical that firms will engage in widespread voluntary public good provision in the absence of a credible threat of regulation. If it were true in general that "it pays to be green," then it would be only a matter of time before managers discovered this and started behaving accordingly. Because the evidence indicates that "it pays to be green" in some ways for some firms in some industries in some countries, but not universally, regulators wanting to see more public good provision need to be ready to use the power of the state to coerce it.

Even more broadly, it is not clear why either activists or regulators should be interested, as a general idea, in granting more discretionary authority over public good provision to corporate managers. Milton Friedman's old (1970) idea that corporate managers who engage in public good provision are acting like unelected government officials is not an academic one. Environmental activists should not assume that corporate managers so empowered will be interested in the same public goods as the environmentalists are. A striking example comes from a recent op-ed in the *Wall Street Journal* (Strassel 2002) about a Montana antienvironmental activist who received $1.5 million from Ford Motor Company for a program called "Provider Pals," under which a primary school class "adopts" a logger or miner, who visits the class periodically to "set the record straight" about rural communities that "often find their livelihoods wrecked by one-sided environmentalism." It should not be totally surprising that a company would choose different externalities to remedy, or different wealth transfers to undertake, than a particular environmentalist would select if given this power.

Lessons for Investors

Whether the literature on environment and corporate social responsibility will be useful to an investor will depend on the investor's objectives and on his or her ideas about market efficiency. At the extreme, true believers in capital market efficiency should not hold anything but the market portfolio, and hence should find unsystematic risk considerations of the sort emphasized in the environmental literature to be uninteresting. By contrast, investors who are interested in maximizing the risk-adjusted expected returns to their portfolios, but who believe that for some reason the marginal investor is ignoring environmental considerations, may find cause in the literature to tilt their portfolios in an environmental direction. This sword, however, can cut either way: if conscientious or fashion-conscious investors dump socially "irresponsible" stocks, these equities will be underpriced. Finally, investors who wish to constrain their portfolios to include only the leaders in the "social responsibility" movement, or to exclude the true "laggards," should find the literature quite interesting in helping them identify the members of those groups—which, as discussed above, is not a straightforward matter—and in showing them what, if any, penalty to expect for their conscientiousness, a topic on which, again as discussed above, the jury is still out.

Lessons for Managers and Executives

It is interesting that the business literature on sustainable development and corporate social responsibility may be of less use to executives and managers than it is to activists and regulators. Managers who are unmoved by moral exhortation and unpersuaded by the "business case for sustainability" will not see much use for the "roadmaps" and checklists that are supposed to help them move in that direction. On the other hand, managers who for one reason or another have decided to engage in more public good provision may benefit from knowing how other firms have approached similar problems.

One point that emerges from the literature is how murky, to those outside a company, are the firm's objectives and constraints. This may appear to present an opportunity for strategic representation of those objectives and constraints. In fact, one function of the literature may be to provide tactical cover for managers who want for their own reasons to engage in greater public good provision. By design or by accident, the literature does tend to obscure questions of motive, blurring Baron's distinction between corporate social responsibility and corporate social performance. This makes it easier for big, well-funded firms to reap reputational benefits for making investments that are sensible in any case, and may encourage activists and regulators to put pressure on less liquid competitors to incur costs.

The difficulty that academics have had in measuring environmental performance also has managerial implications. In a well-designed study of the relationships between financial and environmental performance, Shameek Konar and Mark Cohen (2001) regressed Tobin's q on a number of variables, including toxic emissions per dollar of revenue and number of pending environmental lawsuits. Both environmental variables had the expected sign in the regression; Konar and Cohen, like other scholars, could not show causality. The curious fact is that the two environmental variables were uncorrelated. Jennifer Griffin and John Mahon (1997) found no close relationship between perceptual measures of corporate social responsibility (measures of the company's rank on a Fortune "Most Admired" survey that asked specifically about CSR) and putatively more objective ones (TRI data). These results suggest that the benefits of voluntarism, in the form of enhanced public reputation and a more secure social license to operate, may not follow directly or automatically from investments in improved performance, and that some firms, in the short run at least, may be able to derive such benefits even in the absence of performance improvements.

Lessons for Business Scholars

What sort of research should business scholars be conducting? The answer depends, of course, on their objectives. Business scholars can be seen as economic agents, making production decisions in such a way as to maximize some objective function that presumably includes in its arguments income, but also may include such things as status, intellectual excitement, and the sense of contributing to the solution of private or social problems. Different kinds of outputs will have different payoffs with respect to these different objectives.

In particular, business scholars often engage in both traditional positive scholarship and normative business policy work. Some tension exists between the two roles. In their roles as traditional scholars, business school faculty typically draw on economics and political science in order to explain observable outcomes as arising from optimizing behavior by individuals or by groups that behave like individuals; the audience for this work is likely to be academic. But when the same scholars want to give advice to managers, whether in the classroom, in consulting engagements, in practitioner-oriented papers, or in the business press, they need to have prescriptive recommendations and thus must be assuming that someone somewhere is not behaving optimally. Business scholars need to think about their audiences: Whom would they like to have read their papers? Are they trying to affect behavior? If so, whose?

From a purely scholarly perspective, the approaches that seem most intellectually appealing are those that bring the debate on business and environment closer to conventional economic or organizational scholarship. Formal modeling, statistical tests, and rigorous qualitative analysis of historical cases all can shed light on the behavior of firms or of the managers within them. What seems important, whatever the methodology, is that the work reflect the importance of the economic climate in which the firm operates, the social and regulatory forces that might alter that economic climate, and the organizational dynamics and even individual biases that affect the firm's response.

Turning to research that has normative implications, there may again be trade-offs: the papers that investors find most interesting may have limited appeal for activists, and so on. But I think that at least two streams of research ought to be of considerable interest to activists, regulators, investors, and managers.

First, no matter to which of these groups I belonged, I would be interested in learning more than I currently can discover about what Baron calls "strategic corporate social performance." In particular, I would want to know a lot more about how the provision of environmental public goods affects the two quantities about which managers need to care most deeply: willingness to pay and cost.

For example, despite a lot of general statements of green consumers, we seem to know very little about how the willingness to pay for environmental attributes in products and in production processes varies across product categories, market segments, distributions, nations, and time. It seems clear from a little introspection, for example, that a given individual consumer is likely to have substantially different willingnesses to pay for a particular environmental benefit at different times and in different contexts, just as the willingness to pay for any other product attribute varies across use occasions; and it seems likely that the determinants of these differences would include opportunities to communicate with others about one's status or affiliations, again as it does for other product attributes. If, as Holliday et al. (2002) report, "60% of [British] consumers have looked for products with ethical qualities," but "only 5% do so consistently," an enormous amount of value remains to be captured by firms whose marketers can deepen their understanding of the behavior of the other 55 percent. To make progress on this topic, firm-level data and analysis of CEO speeches may be of limited use; one would need to understand the economics in detail at the product line level.

Similarly, whether I were an investor, a manager, a regulator, or an activist, I would like to know more about the relationships between environmental performance and private production cost. Neither side in the debate opened by Palmer et al. and Porter and van der Linde in 1995 has made any progress in convincing the other. One might ask, as a starting point, whether one can distinguish ex ante between firms that are likely to enhance shareholder value using the environment as a cost-cutting tool and those that are more likely to benefit from other management priorities. One also might look harder at the question of whether savings in labor markets can arise from environmental spending, which seems to be an article of faith for many managers but has not received much systematic exploration. Again, finer-grained data than those at the firm level are likely to be necessary in order to make progress.

Second, I think that investors, managers, regulators, and activists all have a stake in understanding the environmental behavior of firms as a particular example of a far broader class of problems involving the relationships among business and politics and the role of the firm in society. Management of environmental externalities has become an important governmental function. The behavior of the legislatures and administrative bodies that conduct this function, and the behavior of the firms whose opportunities are affected, should be studied in the broader context of business–government relations, as particular examples of far more general phenomena.

To cite just one example, people hold very strong but often conflicting views on the degree to which the pursuit of competitive advantage ought to be socially sanctioned when it impinges on public good provision, income distribution, or the governmental processes that societies put in place to distribute income and provide public goods. For example, the prominent international environmental organization WWF (n.d.) recently wrote:

> One relevant criticism of CSR, both from grassroots organizations and some Friedmanite economists, is that companies should not be proactive in a way that undermines democratic bodies. Here it is important to distinguish between agenda setting and implementation. Companies can be proactive when it comes to implementing environmental goals, but should not engage in agenda-setting lobbying that takes important decisions in society away from democratic institutions and excludes relevant stakeholders.

WWF's idea is that the company's legitimate pursuit of shareholder value does not include involvement in political procedures that shape the rules under which the pursuit of shareholder value is supposed to take place. It seems likely that this view is widely shared among activists, and perhaps among regulators. It is less clear whether it is widely shared among executives and, if not, how social expectations of the company's role in the political process shape the firm's optimal political strategy.

More broadly, as Vogel (1996) notes, students of the environmental behavior of business rarely have related their work to that conducted by other scholars on related topics: "students of environmental policy or economic deregulation are unlikely to be familiar with recent scholarship on business campaign spending" (*159*). Vogel asserts that the study of business–government relations in general will benefit if it places the issues under study in their historical context, exploits cross-national differences to obtain a comparative perspective, and maintains an interdisciplinary approach (*151*). This advice applies to the field of business and environment, and the most promising research exhibits all three of these characteristics.

In fact, one way of bridging the positive and the normative facets of academic work on business and the environment is to note that business scholars, if they can provide an improved understanding of the relationships among firms' behavior, social costs, and the private benefits to the firms' shareholders, can contribute to the convergence of private and social costs that is necessary for continued progress in addressing environmental problems. Business scholars' collective understanding of the relationships among sustainable development, corporate social responsibility, and shareholder value is not yet what it ought to be, but the progress of the last five years or so is encouraging. That progress will continue if researchers keep approaching the question as they would other questions of economics, strategy, and organization, iterating between traditional statistical methods and the analysis of historical cases, and considering failures as well as successes. In this work, the academics will need the help of the firms in sharing disaggregated data and acknowledging that institutional changes beyond selective corporate voluntarism can be socially desirable.

Acknowledgments

For helpful conversations, I would like to thank Antonio Bento, University of California–Santa Barbara; Thomas Dyllick, University of St. Gallen; Aileen Ionescu-Somers, IMD; Dinah Koehler, Wharton; Josh Margolis, Harvard; Amanda Merryman, Harvard; Oliver Salzmann, IMD; Rob Stavins, Harvard; Ulrich Steger, IMD; and Dick Vietor, Harvard.

References

Arora, Seema, and Timothy Cason. 1995. An Experiment in Voluntary Environmental Regulation: Participation in EPA's 33/50 Program. *Journal of Environmental Economics and Management* 28: 271–286.

Baron, David P. 2001. Private Politics, Corporate Social Responsibility, and Integrated Strategy. *Journal of Economics and Management Strategy* 10(1): 7–45.

Baron, David P. 2003. *Business and Its Environment.* 4th ed. Upper Saddle River, New Jersey: Prentice-Hall.

Chouinard, Yvon, and Michael S. Brown. 1997. Going Organic: Converting Patagonia's Cotton Product Line. *Journal of Industrial Ecology* 1(1): 117-130.

Coglianese, Cary, and Jennifer Nash. 2001. Environmental Management Systems and the New Policy Agenda. In *Regulating from the Inside: Can Environmental Management Systems Achieve Policy Goals?* edited by Cary Coglianese and Jennifer Nash. Washington, DC: Resources for the Future.

Dowell, Glen, Stuart Hart, and Bernard Yeung. 2000. Do Corporate Global Environmental Standards Create or Destroy Market Value? *Management Science* 46(8): 1059–1074.

Esty, Daniel C. 2001. A Term's Limits. *Foreign Policy* 126 (September–October): 74–75.

Friedman, Milton. 1970. The Social Responsibility of Business Is to Increase Its Profits. *New York Times Magazine,* September 13.

Froot, Kenneth A., David S. Scharfstein, and Jeremy C. Stein. 1994. A Framework for Risk Management. *Harvard Business Review* (November–December): 91–102.

Griffin, Jennifer J., and John F. Mahon. 1997. The Corporate Social Performance and Corporate Financial Performance Debate: Twenty-Five Years of Incomparable Research. *Business and Society* 36(1): 5–31.

Gunningham, Neil, Robert A. Kagan, and Dorothy Thornton. 2003. *Shades of Green: Business, Regulation, and Environment.* Stanford, CA: Stanford University Press.

Hamilton, Kirk. 2000a. Genuine Saving as a Sustainability Indicator. World Bank Environment Department Paper 22744.

Hamilton, Kirk. 2000b. Sustaining Economic Welfare: Estimating Changes in Per Capita Wealth. World Bank Policy Research Working Paper 2498.

Hart, Stuart L. 1997. Beyond Greening: Strategies for a Sustainable World. *Harvard Business Review* (January–February): 66–76.

Hart, Stuart L., and Mark B. Milstein. 2003. Creating Sustainable Value. *Academy of Management Executive* 17(2): 56–69.

Hartwick, John M. 1977. Intergenerational Equity and the Investing of Rents from Exhaustible Resources. *American Economic Review* 67(5): 972–974.

Henriques, Irene, and Perry Sadorsky. 1996. Determinants of an Environmentally Responsive Firm. *Journal of Environmental Economics and Management* 30(3): 386–395.

Hoffman, Andrew J. 1997. *From Heresy to Dogma: An Institutional History of Corporate Environmentalism.* San Francisco: New Lexington Press/Jossey-Bass.

Holliday, Charles O., Jr., Stephan Schmidheiny, and Philip Watts. 2002. *Walking the Talk: The Business Case for Sustainable Development.* Sheffield, UK: Greenleaf Publishing.

Howard, Jennifer, Jennifer Nash, and John Ehrenfeld. 2000. Standard or Smokescreen? Implementation of a Voluntary Environmental Code. *California Management Review* 42(2):63-82.

King, Andrew A., and Michael J. Lenox. 2001. Does it *Really* Pay to Be Green? An Empirical Study of Firm Environmental and Financial Performance. *Journal of Industrial Ecology* 5(1): 105–116.

Konar, Shameek, and Mark A. Cohen. 2001. Does the Market Value Environmental Performance? *Review of Economics and Statistics* 83(2): 281–289.

Lovins, Amory B., L. Hunter Lovins, and Paul Hawken. 1999. A Road Map for Natural Capitalism. *Harvard Business Review* (May–June): 145–158.

Margolis, Joshua Daniel, and James Patrick Walsh. 2001. *People and Profits? The Search for a Link between a Company's Social and Financial Performance.* Mahwah, NJ: Lawrence Erlbaum Associates.

Maxwell, John W., and Thomas P. Lyon. 2001. An Institutional Analysis of Environmental Voluntary Agreements in the United States. In *Environmental Contracts: Comparative*

Approaches to Regulatory Innovation in the United States and Europe, edited by Eric W. Orts and Kurt Deketelaere. London: Kluwer Law International.

Maxwell, John W., Thomas P. Lyon, and Steven C. Hackett. 2000. Self-Regulation and Social Welfare: The Political Economic of Corporate Environmentalism. *Journal of Law and Economics* 43: 583–617.

Palmer, Karen, Wallace E. Oates, and Paul R. Portney. 1995. Tightening Environmental Standards: The Benefit–Cost or the No-Cost Paradigm? *Journal of Economic Perspectives* 9(4): 119–132. Reprinted in *Economics of the Environment: Selected Readings,* Robert N. Stavins, ed., 4th ed. New York: Norton.

Porter, Michael E., and Claas van der Linde. 1995. Toward a New Conception of the Environment-Competitiveness Relationship. *Journal of Economic Perspectives* 9(4): 97–118. Reprinted in *Economics of the Environment: Selected Readings,* Robert N. Stavins, ed., 4th ed. New York: Norton.

Reinhardt, Forest. 2000. *Down to Earth: Applying Business Principles to Environmental Management.* Boston: Harvard Business School Press.

Reinhardt, Forest. 2003. Tests for Sustainability. In *The Global Competitiveness Report 2002–2003,* edited by Peter K. Cornelius. New York: Oxford University Press for the World Economic Forum.

Salop, Steven C., and David T. Scheffman. 1983. Raising Rivals' Costs. *American Economic Review* 73(2): 267–271.

Solow, Robert M. 1974. Intergenerational Equity and Exhaustible Resources. *Review of Economic Studies*: 29–45.

Solow, Robert M. 1991. Sustainability: An Economist's Perspective. J. Seward Johnson Lecture to the Marine Policy Center, Woods Hole Oceanographic Institution, June 14, 1991. Reprinted in *Economics of the Environment: Selected Readings,* Robert N. Stavins, ed., 4th ed. New York: Norton.

Strassel, Kimberley. 2002. Hug a Logger, Not a Tree. Op-ed. *Wall Street Journal,* May 23.

SustainAbility and IFC (International Finance Corporation). 2002. *Developing Value: The Business Case for Sustainability in Emerging Markets.* London: SustainAbility.

SustainAbility and UNEP (United Nations Environment Program). 2001. *Buried Treasure: Uncovering the Business Case for Corporate Sustainability.* London: SustainAbility and United Nations Environment Program.

Vincent, Jeffrey. 2001. Are Greener National Accounts Better? CID Working Paper no. 63. Cambridge, MA: Harvard Center for International Development.

Vogel, David J. 1996. The Study of Business and Politics. *California Management Review* 38(3): 146–165.

WBCSD (World Business Council for Sustainable Development). 2000. *Corporate Social Responsibility: Making Good Business Sense.* Geneva.

Weitzman, Martin. 1976. On the Welfare Significance of National Product in a Dynamic Economy. *Quarterly Journal of Economics* 90(1): 156–162.

World Commission on Environment and Development. 1987. *Our Common Future.* New York: Oxford University Press.

Worldwide Fund for Nature (WWF). n.d. WWF Discussion Paper; Corporate Social Responsibility, CSR—An Overview. N.p.

Comment on Reinhardt

Ethics, Risk, and the Environment in Corporate Responsibility

Eric W. Orts

Forest Reinhardt provides an excellent survey of some of the recent business literature concerning corporate governance and environmental management. His review is especially strong in providing a critical overview of work that has focused on whether and when "it pays to be green" as measured primarily in terms of shareholder value. A unifying theme of his inquiry is that the relevant business research looks to the fundamental question of "the business case" for environmental management. Reinhardt's particular interest is in "the application of traditional business principles" to thinking about environmental problems that can lead to financial payoffs or, in other words, "win–win" solutions.[1] He also provides a useful compilation of studies that examine various aspects of environmental management, including case studies of attempts to differentiate and market environmentally friendly products and services. Even if these examples do not aggregate to large-scale observable results, they are important and often persuasive.[2]

This direction in both the research and teaching of environmental management is very important, because it answers the perennial call in business circles for focusing primarily on profits and shareholder value. I do not think, however, that Reinhardt's survey gets to the root of the idea of corporate social responsibil-

1. See also Forest L. Reinhardt, "Bringing the Environment Down to Earth," in *Harvard Business Review on Business and the Environment* (2000a), p. 36.

2. A major contribution in this field is Forest Reinhardt's own recent book. Forest L. Reinhardt, *Down to Earth: Applying Business Principles to Environmental Management* (2000). For an argument in a similar spirit maintaining that social and financial values are mutually reinforcing, see Lynn Sharp Paine, *Value Shift: Why Companies Must Merge Social and Financial Imperatives to Achieve Superior Performance* (2003).

ity as it has been traditionally understood. In addition, empirical evidence to support some of Reinhardt's arguments is sparse. "Win–win" solutions are very persuasive when they occur, but they may not be as prevalent as Reinhardt and some other scholars in this field may like to believe. Lastly, Reinhardt does not articulate an agenda for businesses that would point them toward strategies to achieve socially desirable outcomes when difficult choices cannot be rendered into the economics of costs and benefits. Additional ethical arguments for proactive environmental management can and should be made to persuade firms to commit themselves to a view of corporate responsibility that extends beyond "win–win" cases. This approach recommends self-imposed moral duties in the environmental area that correspond to traditional ideas of corporate social responsibility applied in other areas. In my view, the concept of corporate responsibility should updated to include the environment as an important element—not because business may or may not find doing so in its best economic interests, but because there is a compelling ethical need for business to respond to the increasingly severe environmental risks that its activities impose on society.

Ethics, Risk, and Corporate Social Responsibility

Corporate social responsibility (CSR) is an ethical idea with a long pedigree, stretching back at least to the time of Walther Rathenau and the Weimar Republic in Germany. In this received tradition, CSR means that the moral responsibilities of business organizations cannot be reduced to an economic objective of profits. Employees, for example, must be treated with the basic respect and dignity that human beings deserve and not merely as "cogs in the wheel" of business machinery or simply as an abstract "factor of production." The history of the idea has been sketched briefly elsewhere,[3] and it is broadly discussed in standard business school texts and the academic literature.[4] Reinhardt discusses only a small sample of this literature, which would indicate that he entertains a different concept of corporate responsibility than has been traditionally understood.

3. See, e.g., Eric W. Orts, "A North American Perspective on Stakeholder Management Theory," in *Perspectives on Company Law: 2* (Fiona Macmillan Patfield, ed., 1997), pp. 168–71. See also Eric W. Orts, "From Corporate Social Responsibility to Global Citizenship," in *The INSEAD-Wharton Alliance on Globalizing* (Hubert Gatignon and John Kimberly, eds., 2004), pp. 331–352.

4. Standard texts include Archie B. Carroll and Ann K. Buchholtz, *Business and Society: Ethics and Stakeholder Management* (5th ed. 2002); William C. Frederick, James E. Post, and Keith Davis, *Business in Society: Corporate Strategy, Public Policy, Ethics* (7th ed. 1992); *Ethical Issues in Business: A Philosophical Approach* (Thomas Donaldson et al., eds., 7th ed. 2002). Samples of different versions of corporate social responsibility in the literature appear in *Harvard Business Review of Corporate Responsibility* (2003); *Corporate Governance and Directors' Liabilities: Legal, Economic, and Sociological Analyses on Corporate Social Responsibility* (Klaus J. Hopt and Gunther Teubner, eds., 1985). For a provocative recent book on the topic, see Lawrence E. Mitchell, *Corporate Irresponsibility: America's Newest Export* (2001).

Reinhardt refers to David Baron's distinction between "corporate social responsibility" (CSR) and "corporate social performance" (CSP).[5] For Baron, CSR involves the redistribution of a firm's wealth to public goods motivated by normative, that is, ethical, principles. True CSR therefore is founded on an ethical commitment to the public good. Mere "performance," or CSP, on the other hand, refers only to a redistributive effect or public service without regard to its motivation. CSP thus may involve only the appearance of an ethical motivation. CSP may be driven purely by profits—for example, by the calculation that a good public reputation may provide economic benefits in the form of reduced regulation. Understanding the distinction between CSR and CSP allows for an appreciation of the worst possible approach of "strategic CSR." Here the firm follows a profit-maximizing strategy motivated entirely by self-interest, and CSR is used literally as a false label or "front" to blunt outside social criticism of the firm's behavior. Strategic CSR is devoid of any normative principles other than economic wealth maximization. It can include outright deception. For example, "greenwashing" charges are properly classified as allegations of strategic CSR. A "greenwashing" firm pretends to be motivated by larger ethical concerns, such as mitigating its adverse effects on the natural environment, when in fact it is driven only by profits.

Certainly, situations in which a business can cut costs, save money, and exploit "green consumer" niche markets by taking the environment seriously are important. However, these situations do not pose the most difficult ethical questions; they are "easy cases" for a paradigm that assumes that profits and shareholder gain are the exclusive objectives of any business.[6] The questions on which Reinhardt focuses most of his attention—namely, whether and when "it pays to be green"—are not really about corporate social responsibility. Instead, problems of true CSR enter the analysis exactly when the profit motive—at least, in the relevant time horizon—runs out. For example, how does a manager make a decision about pursuing an internal project that might cut costs and increase profits (such as a life-cycle analysis for a proposed product that identifies savings through recycling or waste reduction), but also might not (if the costs of the life-cycle analysis exceed the gains)? An ethical inclination toward making choices about competing projects may prompt firms to accept financial risks in the face of uncertain returns and go ahead with these kinds of projects. Another example reinforces the point. A firm's decision to adopt internal measures to reduce car-

5. David P. Baron, "Private Politics, Corporate Social Responsibility, and Integrated Strategy," J. Econ. Mgmt. Strategy 10: 7–45 (2001).

6. This is not to say that learning to "make the business case" for environmental improvements is not important, and a critical skill for business students and others who may be faced with a skeptic more favorably disposed to initiatives that also can lead to improvements along traditional lines (e.g., reduced costs or otherwise higher profits). Just as environmentalists must learn the language of economics to make the best case in court, so environmentalists who want to influence business must learn to argue in terms of the economic costs and benefits to the business.

bon emissions might incorporate a longer time horizon than usual in order to justify making such an investment. In the end, decisions of this kind about the relevant time horizon may rely as much on the *ethical* argument that climate change is a large and important social issue that should be addressed as on economic arguments that *eventually* climate change is likely to have financial consequences.

On another dimension, Reinhardt correctly emphasizes that businesses should adopt rational environmental risk-management strategies from an economic perspective. For example, it is helpful to estimate the probability of a major chemical accident in one's line of business, because such an accident will impose immediate and longer-term costs on the firm: loss of property and productivity, as well as compensation for workers and legal liability for serious injuries or deaths caused by the accident in the surrounding community. Reputational costs to the firm—which may result in reduced sales or even a threat to its "social license to operate"—are also likely to be important.

One problem with this line of argument that makes economic motivation the key explanation for environmental risk management, however, is the lack of strong empirical evidence supporting such a connection. To date, some correlations have been found between environmental performance (variously defined) and financial performance (also variously defined). However, most of this empirical work suffers from serious methodological flaws, which undermine the conclusions of statistical relationships.[7] A statistical relationship between better environmental performance and better financial performance has been difficult to demonstrate, for several important reasons omitted in Reinhardt's review. First, it is not at all clear that social or green investors can exert enough influence to lower the cost of capital for firms characterized by their policies of corporate social responsibility, however these are defined or understood. On the contrary, a recent study by some Wharton School colleagues shows that social investors can expect to pay a financial penalty for green mutual funds, as compared with unconstrained investment strategies, thus casting considerable doubt on whether it very often "pays to be green" for shareholders.[8] Second, Reinhardt suggests that environmental risk management can impact the expected value of firms and variance of cash flows. However, unless environmental risks are systemic—that is, correlated with fundamental market drivers and affecting major portions of the economy—long-term investors who are fully diversified can safely ignore these risks. Reinhardt points to no convincing empirical research to support the claim that voluntary risk-reduction efforts directly affect cash flows or overall equity returns.

---

7. Dinah A. Koehler, "Capital Markets and Corporate Environmental Performance: Evidence of a Link?" Wharton School working paper (2004).

8. Christopher C. Geczy, Robert F. Stambaugh, and David Levin, "Investing in Socially Responsible Mutual Funds," Wharton School working paper (May 26, 2003), http://finance.wharton.upenn.edu/~geczy/workingpapers.htm (accessed September 1, 2004).

Fortunately, economic arguments are not the only approach available for a business that may wish to make investments in environmental risk management. Risk management is not simply a matter of economics, but also ethics. Many firms impose social risks that can often result in serious injuries and deaths, as well as long-lasting and unquantifiable environmental damage. Consider, for example, the now well-known examples of the widespread social damage caused by exposure to lead, asbestos, and tobacco smoke. These and other cases suggest that there is an ethical dimension to environmental risk management. Understanding that an aspect of one's business operations runs a significant risk of harming employees, neighbors, or the surrounding natural environment leads to a moral as well as an economic duty to manage the risk appropriately and rationally. Union Carbide's tragic chemical accident in Bhopal or the Exxon *Valdez* oil spill come to mind as prominent specific examples. These single accidents triggered ripple effects across entire industry sectors. They also resulted in major regulatory interventions: in the case of chemicals, the adoption of the Toxics Release Inventory and requirements for Risk Management Plans under the Clean Air Act; and in the oil industry, a mandate requiring double-hulled tankers and new insurance and liability rules. Since Bhopal, many chemical companies have decided to apply global safety standards to all facilities, regardless of the possibility of more lenient local laws.[9]

Arguably, an economic calculus drives these decisions, but it is more likely that moral considerations predominate. Fundamental moral theories agree that a human life in India is not worth less than a life in the United States. Moral arguments support the view that a company should treat all of its employees similarly—at least at the level of general policy—including the adoption of uniform safety standards for similar plants. One does not need to read Kant or invoke religious versions of the "golden rule" to agree that a principle of basic respect for human beings would justify treating employees similarly with respect to subjecting them to the risks of serious injury or death. The argument applies with even more force to risks imposed on a surrounding community, because, unlike employees, those living there are not in a position to consent to increased risk.[10]

9. The CEO of a major chemical firm relocating a plant from the United States to Mexico has said, against economic arguments, that the same safety standards should apply in Mexico, despite the legal ability to relax them. Other scholars support the general claim that large companies in fact often make decisions to apply uniform environmental and safely standards to operations throughout the world, despite differences in regulations.

10. Some object that one would not feel similarly about paying Mexican employees less than those in U.S. facilities. I agree, but I do not think that this argument is the same as deciding to impose significantly different risks of accidental death on one's employees. One might say that the different wage structures in different countries are a function of many different factors beyond a company's control. Perhaps there is a moral argument that companies should pay all similarly situated employees comparatively equal wages. At the same time, one might think it is equally plausible that a Mexican employee would agree to a lower wage than an American employee in terms of the benefit to his or her life prospects. It is difficult to see why a Mexican employee would agree, however, to a higher

It does not follow, of course, that a business has an ethical responsibility to reduce the risks that it imposes on life and limb, flora and fauna, to zero. Such radical risk aversion is not the choice that contemporary society as a whole has adopted, and it also is not a position required of any particular business. Rather, the point is that a business must take ethical responsibility for a serious deliberative approach to managing the serious environmental risks that it imposes on its employees, society generally, and the natural environment. Law presents one standard by which risk management may be measured. The law permits a specified level of certain kinds of air and water pollution known to cause some harm to public health, because a political decision has been made to allow this level of risk. Such legal permissions, however, are not the end of the matter.

Business, Government, and the Environment: Beyond Milton Friedman

Perhaps Forest Reinhardt treats corporate social responsibility only in economic terms because he is, at least implicitly and perhaps unconsciously, a disciple of Milton Friedman.[11] For Friedman, there is an easy answer to the problem of determining what environmental risks are morally acceptable. He argues that the social good is maximized when business "sticks to its knitting" and focuses strictly on profits, leaving to government the task of regulating public goods, including noneconomic risks. In his words, corporate executives have the responsibility "to make as much money as possible while conforming to the basic rules of the society, both those embodied in law and those embodied in ethical custom."[12] For Friedman, going beyond the profit motive is dangerous nonsense at best, and at worst, it is socialism in a business suit.

Notice, however, that at the same time that Friedman exhorts business to focus on making "as much money as possible," he admits that business should follow the law, presumably on moral grounds, and conform to other unspecified rules of "ethical custom." One assumes that Friedman would not condone, for example, a business strategy to conduct a cost–benefit analysis about whether to comply with an environmental law restricting pollution. As for "ethical custom," Friedman supplies little further guidance. If he understands it to mean simply the observed behavior and opinions of other corporate executives, then he is enthralled in a circular logic. Ethical obligations cannot depend merely on the

risk of violent death than his or her American counterpart; and even if some employees would agree to such a higher risk, it would seem problematic for a company to agree to impose higher risks of death, rather than lower wages, on different groups of people.

11. Although Reinhardt does not directly rely on Friedman, his assumptions about corporate social responsibility seem to be broadly consistent with Friedman's approach and would explain his amoral approach to the topic.

12. Milton Friedman, "The Social Responsibility of Business Is to Increase Its Profits," N.Y. Times Magazine (1970).

happenstance of what currently passes for acceptable. The ethical obligations of business should instead be subject to sustained analytical scrutiny.[13]

The exceptions of legal and ethical constraints on the profit motive mentioned by Friedman are recognized in the corporate governance principles of the American Law Institute (ALI). The main objective of a business corporation, according to the ALI, is "enhancing corporate profits and shareholder gain." "Even if corporate profit and shareholder gain are not thereby enhanced," however, the ALI recognizes a mandatory obligation "to act within the boundaries of the law" and permits a company to "take into account ethical obligations that are reasonably regarded as appropriate to the responsible conduct of business."[14] The current law of the United States, then, confirms a liberal extension of Friedman's version of corporate social responsibility. The ALI's restatement of the law makes room for corporate social responsibility in its true sense: that is, business behavior that may "sacrifice" profits (in Einer Elhauge's terms) in pursuit of other goals deemed by society as worthy, such as environmental sustainability.[15] What exactly "environmental sustainability" may mean in the business context is notoriously uncertain. However, it would seem to fit easily in the larger agenda of CSR.

One of the most deceptively attractive aspects of Friedman's rejection of CSR is that his position allows corporate executives conveniently to ignore the effects that their operations may have on society and the natural environment, except to the extent that these effects are translated into formal economic and legal constraints. According to this view, business should trust the liberal political processes of democracy and accept legislation that restricts the scope of its activ-

13. One possibility in the environmental context would be to apply ethical standards of sustainability to evaluate business practices. Specifically, Reinhardt refers to definitions of "sustainable development" that focus on total national wealth as a function of natural plus human-made capital, where gains from exploitation of natural capital are offset by investment in human-made capital for net social gain–national income. It is important to note that this economic definition of sustainability at the national level is not espoused by the World Business Council for Sustainable Development (WBCSD), though it sounds very much like Baron's definition of CSR. WBCSD treats CSR as a *voluntary* transfer of wealth for the purposes of sustainable development, which is positively connected to business advantage, i.e., CSP or even strategic CSR. A major problem with this approach is that the environmental effectiveness and economic efficiency of voluntary approaches are often questionable.

14. American Law Institute, *Principles of Corporate Governance: Analysis and Recommendations* (1994), § 2.01. The ethical exceptions to the profit motive also include the corporate power to "devote a reasonable amount of resources to public welfare, humanitarian, educational, and philanthropic purposes." Id.

15. Einer Elhauge, in "Corporate Managers' Operational Discretion to Sacrifice Corporate Profits in the Public Interest" in this volume. Note that even corporate executives who wish to "sacrifice" profits to some limited extent in order to advance other goals of social responsibility face economic and legal constraints in markets for goods and services and in the legal structures of corporate governance that give shareholders significant powers to monitor, and in some cases, even to replace corporate managers.

ity. This view might be viable if one could assume that business would not involve itself actively in the political process. Yet this assumption is contradicted by an everyday understanding of the significant role that business actually plays in direct lobbying and political campaigns, a concern that Reinhardt, to his credit, recognizes. If business aims to influence politics so as to maximize profits, then the Friedmanesque argument that one can depend on the liberal democratic process to constrain business behavior becomes naive at best, or at worst, hypocritical. Recognizing the inevitability of business involvement in the political system might lead instead to a consideration of the ethical limits that a business should impose on itself or, more realistically, limits that government should impose on business through legal restrictions.

In the regulatory context, the lingua franca of environmental risk management returns in a different form. Well-informed environmental risk assessment is important not only in terms of economic calculations, but also in terms of public debates about levels of acceptable social risk.[16] When lobbying the government, it is ethically untenable for a business to ignore public concerns about environmental risks. Reinhardt expresses a general concern that giving firms discretionary authority over the provision of public goods will not lead to results that are in the interest of the public or environmentalists. But he seems blind to the troubling problem that business often will tend to advance its own economic interests by obstructing or deforming the adoption of rigorous legislative reform designed for the public good. Ethical arguments here can provide some strong grounds at least for restricting inappropriate use of business power, if not for a more positive channeling of the political influence of business toward public ends.

A political model of CSR in this context suggests several basic principles. One is that business should guard itself from an overly self-interested view of legislation. An ethical rule of thumb in lobbying would be for a business to restrict itself to making good-faith arguments on behalf of proposed legislation, or lack of legislation, in terms of the public interest as well as its own interests.[17] A second principle is that business should take a public-interested perspective when it becomes involved in the political and legislative process. Perhaps the best manner by which to formulate a broad perspective is to engage in conversations with other interest groups. This principle would favor a "collaborative" approach rather than a partisan "us against them" perspective. In the environmental context, for exam-

16. For good examples of discussions of risk assessment in the public arena, see Stephen Breyer, *Breaking the Vicious Circle: Toward Effective Risk Regulation* (1993); *Risk versus Risk: Tradeoffs in Protecting Health and the Environment* (John D. Graham and Jonathan Baert Wiener, eds., 1997).

17. To take an example from a tangentially related field, a company in the business of supplying military weapons should tread very carefully in matters of foreign policy for ethical reasons. War profiteering cannot be ethically justified. For extensions of this argument, see Eric W. Orts, "War and the Business Corporation," Vanderbilt J. Transnational Law 35: 549–84 (2002).

ple, a business often may want to consult with representatives of environmental groups when formulating positions on different issues.[18] A third basic principle is that corporate social responsibility requires honesty as a prerequisite. Whatever position or approach a business may decide to adopt, it should be forthright and direct. Better an honest "dirty" business than a lying "greenwashing" one.

Updating Corporate Social Responsibility to Include the Natural Environment

Sustainability is a vital concept that is not easy to define or explain. Much work remains to be done to integrate environmental considerations into effective business practices, drawing on different disciplines, including ethics, economic analysis, and risk assessment and management. Increasingly, however, business representatives agree that the most important environmental issues in our time call for careful thought and new ideas for proper management. Sustainability in business has come to be an ethical imperative, at least in theory. Translating this idea into the management of businesses as part of society in a manner that leaves future generations as well off as the present generation, perhaps even better off, is an admirable moral as well as economic objective.[19]

A couple of quotations from Paul Hawken's influential 1993 book, *The Ecology of Commerce*, are sufficient to give a flavor of the sense of urgency that some have expressed on this issue:

> Quite simply, our business practices are destroying life on earth. Given current corporate practices, not one wildlife preserve, wilderness, or indigenous culture will survive the global market economy.... There is no polite way to say that business is destroying the world.[20]

> Many companies today no longer accept the maxim that the business of business is business. Their new premise is simple: Corporations, because they are the dominant institutions on the planet, must squarely address the social and environmental issues that afflict humankind.[21]

---

18. A proponent of "collaborative" approaches to politics and legislation along these lines is Jody Freeman. See, e.g., Jody Freeman, "Collaborative Governance in the Administrative State," UCLA L. Rev. 45: 1–100 (1997); Jody Freeman, "The Private Role in Public Governance," NYU L. Rev. 75: 543–675 (2000). A related approach is one that recommends regulatory contracts as an alternative model worthy of consideration for some kinds of problems. See, e.g., *Environmental Contracts: Comparative Approaches to Regulatory Innovation in the United States and Europe* (Eric W. Orts and Kurt Deketelaere, eds., 2001). For the general argument for moral deliberation and open discussion of divisive issues in modern democratic politics, see Amy Gutmann and Dennis Thompson, *Democracy and Disagreement* (1996).

19. This formulation is an adaptation of the classic, though admittedly vague, definition first given in the Brundtland report.

20. Paul Hawken, *The Ecology of Commerce: A Declaration of Sustainability* (1993), p. 3.

21. Id., p. xiii.

Business leaders who have been convinced of this perspective are making strong ethical claims that corporate social responsibility should be expanded to include a primary focus on the natural environment. Unfortunately, they too often succumb to temptation and reduce their commitment to embracing only the easy economic arguments for CSR and sustainability.

Some work has begun in business schools along the lines of making coherent arguments in favor of an expanded notion of CSR to include the natural environment, and Forest Reinhardt provides a reliable guide to much of it. One area of research that eludes his search appears in some recent debates about the scope of "stakeholder theory" in management. Some scholars have argued that existing stakeholder theories can be usefully expanded to include the natural environment as a legitimate "stakeholder."[22] Others have rejected this approach as untenable.[23] The existence of the debate itself indicates movement within business schools in thinking about how the environment fits into received views of CSR. Much work remains to be done in this area of inquiry both theoretically and practically, but work in this direction at least has begun.

By way of conclusion, two different possible directions might be outlined for how an ethically rigorous development of what might be called "environmental corporate responsibility" might proceed. These two directions correspond to the two most influential approaches to ethics in philosophy.

1. *Deontological constraints and duties.* The first approach focuses on the ethical constraints and duties that a business has to refrain from wrongfully harming other human beings and the natural environment in its operations. Specification of these duties is not easy, but some basic principles are relatively uncontroversial. There is a positive duty along Kantian lines, for example, to treat people with basic respect. This duty applies to business decisions with as much force as to other areas of human endeavor. The example of environmental risks of accidents that may involve deaths and serious injury is again useful. A strong deontological argument supports a multinational corporation's decision to employ global safety standards that would, in principle, respect the lives of all its employees and other affected people or perhaps even future generations equally. Another example relates to charges of corporate "greenwashing." A deontological perspective condemns lying as wrong in most circumstances, "greenwashing" included.

   A more difficult dimension concerns what, if any, ethical duties owe directly to the natural environment in its own terms. One might speculate, for example, that some aesthetic or other foundational duty applies to oil

22. See, e.g., Mark Starik, "Should Trees Have Managerial Standing? Toward Stakeholder Status for Non-Human Nature," J. Business Ethics 14: 207 (1995); Robert A. Phillips and Joel Reichart, "The Environment as Stakeholder? A Fairness-Based Approach," J. Business Ethics 23: 185 (2000).

23. Eric W. Orts and Alan Strudler, "The Ethical and Environmental Limits of Stakeholder Theory," J. Business Ethics 12: 215 (2002).

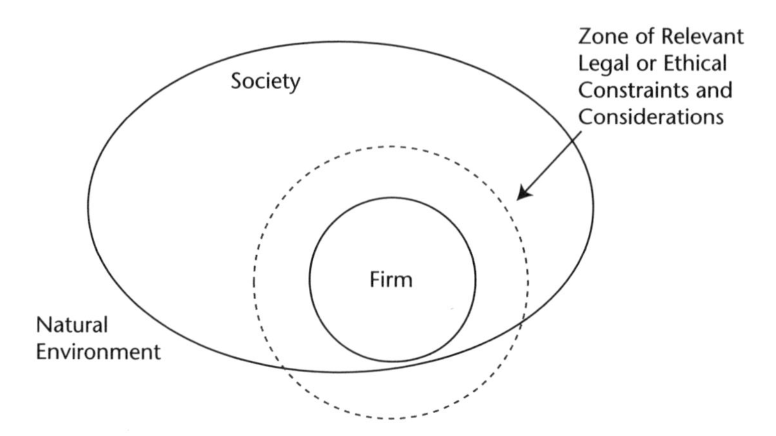

FIGURE 1. *Corporate Social Responsibility and Its Limits*

companies such as Exxon in taking responsibility for the reduction of risks of major oil spills, especially when routes threaten pristine wilderness areas.

2. *Utilitarian ethics.* Another common ethical worldview relies on a utilitarian view in which business plays a role in maximizing social welfare. In this view, if business can add to the social welfare outside of its wealth-maximizing role, it should do so. One method to evaluate this would be social cost–benefit analysis as developed, for example, by administrative agencies such as the U.S. Environmental Protection Agency (EPA). Business should not usurp or improperly influence the political process by which social welfare measures are promoted through law. One might argue here for a principle of "separation of business and state" analogous to the constitutional "separation of church and state." Regulation of lobbying and campaign finance participation is a means toward this end. In more general terms, the business ethics of environmental management should focus on the public good as well as the self-interest of an individual business. This approach is consistent with the idea expressed above that business should engage in public debate in terms of the public interest, and not just strategically or hypocritically, but honestly.

These philosophical approaches to the problem of environmental corporate responsibility seem the most promising. Figure 1 illustrates how—in either a deontological or utilitarian theory—adding the natural environment to CSR can change the focus of a business. Firms cannot, of course, be expected to turn themselves into not-for-profit organizations or to solve the world's problems simply because governments have proven unequal to the task. Instead, a "zone of relevant legal and ethical constraints and considerations" for each particular business may be imagined to extend into the broader society and the natural environment beyond the boundaries of the firm. The limits of CSR, indicated by the dotted line in Figure 1, are variable and often unclear, as are the boundaries of the firm itself.[24] But the dotted line represents areas in which a business and its activities and operations touch most directly on larger social and environmental issues. The fact that a business exists both in society and in the natural environment means that moral responsibility for larger effects cannot be avoided. When larger issues impinge closely on business activity, these issues must be taken into account and rationally managed.

Social demands for corporate social responsibility to include the natural environment are increasing. Forest Reinhardt is correct to observe that institutional changes in this direction may be socially desirable from an economic perspective and driven in part by economic forces. But further work in developing the principles of an environmentally sustainable version of corporate responsibility will involve a number of disciplines: not only economics, but also the science of risk assessment and management, political and legal theories of the proper scope of business activities, and applied business ethics.

Acknowledgment

I would like to thank and especially recognize Dinah A. Koehler, a postdoctoral fellow and administrative director of the Environmental Management Program at Wharton, for her assistance in preparing this comment. Many of the insights expressed here about the relationships among risk, environmental management, and financial performance are hers. Any errors, however, are mine alone.

References

American Law Institute (ALI). 1994. *Principles of Corporate Governance: Analysis and Recommendations*. Philadelphia: ALI.

Baron, David P. 2001. Private Politics, Corporate Social Responsibility, and Integrated Strategy. *Journal of Economics and Management Strategy* 10(1): 7–45.

Breyer, Stephen. 1993. *Breaking the Vicious Circle: Toward Effective Risk Regulation*. Cambridge, MA: Harvard University Press.

Carroll, Archie B., and Ann K. Buchholtz. 2002. *Business and Society: Ethics and Stakeholder Management*. 5th ed. Cincinnati: Thomson South-Western.

Donaldson, Thomas et al. 2002. *Ethical Issues in Business: A Philosophical Approach*, 7th ed. Upper Saddle River, NJ: Prentice Hall.

Frederick, William C., James E. Post, and Keith Davis. 1992. *Business in Society: Corporate Strategy, Public Policy, Ethics*. 7th ed. New York: McGraw-Hill College.

Freeman, Jody. 1997. Collaborative Governance in the Administrative State. *UCLA Law Review* 45(1): 1–100

———. 2000. The Private Role in Public Governance. *NYU Law Review* 75(2): 543–675.

Friedman, Milton. 1970. The Social Responsibility of Business Is to Increase Its Profits. *New York Times Magazine*, Sept. 13, 1970.

---

24. For at least my own conception of the shifting legal and economic boundaries of firms, see Eric W. Orts, "Shirking and Sharking: A Legal Theory of the Firm," Yale L. & Pol'y Rev. 16: 265 (1998).

Geczy, Christopher C., Robert F. Stambaugh, and David Levin. 2003. Investing in Socially Responsible Mutual Funds. Wharton School working paper.

Graham, John D., and Jonathan Baert Wiener (eds.). 1997. *Risk versus Risk: Tradeoffs in Protecting Health and the Environment.* Cambridge, MA: Harvard University Press.

Gutmann, Amy, and Dennis Thompson. 1996. *Democracy and Disagreement.* Cambridge, MA: Belknap Press.

Harvard Business Review. 2003. *Harvard Business Review of Corporate Responsibility.* Boston: Harvard Business School Press.

Hawken, Paul. 1993. *The Ecology of Commerce: A Declaration of Sustainability.* New York: Harpercollins.

Hopt, Klaus J., and Gunther Teubner (eds.). 1985. *Corporate Governance and Directors' Liabilities: Legal, Economic, and Sociological Analyses on Corporate Social Responsibility.* Berlin: de Gruyter.

Koehler, Dinah A. 2004. Capital Markets and Corporate Environmental Performance: Evidence of a Link? Wharton School working paper.

Mitchell, Lawrence E. 2001. *Corporate Irresponsibility: America's Newest Export.* New Haven: Yale University Press.

Orts, Eric W. 1997. A North American Perspective on Stakeholder Management Theory. In *Perspectives on Company Law*, edited by Fiona Macmillan Patfield. New York: Kluwer Law International.

———. 1998. Shirking and Sharking: A Legal Theory of the Firm. *Yale Law and Policy Review* 16(2): 265–329.

———. 2002. War and the Business Corporation. *Vanderbilt Journal of Transnational Law* 35: 549–84.

———. 2004. From Corporate Social Responsibility to Global Citizenship. In *The INSEAD-Wharton Alliance on Globalizing,* edited by Hubert Gatignon and John Kimberly. Cambridge, UK: Cambridge University Press.

Orts, Eric W., and Kurt Deketelaere (eds.). 2001. *Environmental Contracts: Comparative Approaches to Regulatory Innovation in the United States and Europe.* New York: Kluwer Law International.

Orts, Eric W., and Alan Strudler. 2002. The Ethical and Environmental Limits of Stakeholder Theory. *Journal of Business Ethics* 12: 215–233.

Paine, Lynn Sharp. 2003. *Value Shift: Why Companies Must Merge Social and Financial Imperatives to Achieve Superior Performance.* New York: McGraw-Hill.

Phillips, Robert A., and Joel Reichart. 2000. The Environment as Stakeholder? A Fairness-Based Approach. *Journal of Business Ethics* 23(2): 185–197.

Reinhardt, Forest L. 2000a. Bringing the Environment Down to Earth. In *Harvard Business Review on Business and the Environment.* Boston: Harvard Business School Press.

———. 2000b. *Down to Earth: Applying Business Principles to Environmental Management.* Boston: Harvard Business School Press.

Starik, Mark. 1995. Should Trees Have Managerial Standing? Toward Stakeholder Status for Non-Human Nature. *Journal of Business Ethics* 14(3): 207.

Comment on Reinhardt

Opportunities for and Limitations of Corporate Environmentalism

David J. Vogel

Forest Reinhardt addresses a number of important issues. One of the most important has to do with the potential for improved environmental management strategies to enhance economic performance. He critically examines three potential ways in which firms can gain competitive advantages from being "greener": namely, by product differentiation, lowering production costs, and reducing financial and political risks. He correctly notes the limitations of the first of these: at least in the United States, there are remarkably few documented cases of successful ecomarketing. Notwithstanding their responses to public-opinion polls, few consumers appear willing to pay a premium for products that are produced in more environmentally responsible ways—unless such products also provide them with direct benefits, such as organic food.

The situation is somewhat different in Europe, in part because quasipublic ecolabeling schemes provide consumers with credible information about the environmental impact of particular corporate practices. But even in Europe, successful cases of ecomarketing are primarily confined to northern Europe and to a relatively small number of highly visible products, such as chlorine-free paper. In some cases, retailers on both sides of the Atlantic have been pressured successfully to offer green certified products, most notably wood products certified by the Forest Stewardship Council or similar bodies, but here again, such cases are unusual.

I think, however, that Reinhardt may insufficiently appreciate the potential of cost savings from greener production strategies. In principle, such savings should not exist. A well-run firm should always be searching for and adopting cost-savings measures; it should not require any particular public pressures or social commitment for managers to identify additional ways of reducing costs. And presumably some portion of these cost savings also will reduce the firm's environmental impact, such as by reducing energy consumption or using fewer raw materials.

But few such well-run firms exist in the real world. Firms are always operating below the optimum level of efficiency. Managers are invariably myopic; they typically lack adequate information about new investment opportunities, promising technologies, global market opportunities, new production methods, and the like. It therefore should not be surprising that as their awareness of environmental issues has increased, they have become more aware of new opportunities to benefit shareholders by reducing their firm's environmental "footprint." There is no reason to think that the number of such opportunities has been, or ever will be, exhausted.

This does not make environmental management a unique case. On the contrary, it makes it similar to every other kind of business opportunity. Consider, for example, another challenge many firms face: namely, encouraging highly skilled and trained female managers to assume progressively greater responsibilities, rather than take early retirement or work part-time. If a firm decided to make a serious effort to better utilize the talents of its female managerial employees, is it not likely that it would discover cost-effective ways of doing so? And would not the same also be true if the firm were to make a more systematic effort to identify and purchase promising technologies developed by other firms?

In short, an infinite number of five-euro bills are waiting to be picked up. And some of these also will produce environmental benefits.

Yet at the same time, we also need to recognize the limitations of corporate environmentalism. Stuart Hart, Amory Lovins, and scores of other writers and speakers on "sustainable management" make two arguments. The first is that our environmental problems are extremely serious. The second is that managers, operating within a market economy, can address these problems effectively—if only they were sufficiently enlightened and creative.

I doubt that either of these claims is true. But both cannot be. If our environmental problems were considered modest, then it is entirely reasonable that innovative management practices could adequately address many of them. We could then look forward to an endless series of marginal improvements in corporate environmental performance, and environmental quality would steadily improve. But if our environmental problems are indeed serious—as virtually every writer on corporate environmentalism claims—then relying on the private sector to ameliorate them is not only naive, but also irresponsible.

The major shortcoming of virtually all the voluminous writings on corporate environmentalism (I must have at least two score such books in my office, and new ones keep arriving regularly) is their lack of attention to public policy. For if we do face critical environmental challenges, they cannot be addressed without more and better government regulation. Yet the need for such regulation is ignored in this literature. Indeed, most writers on environmental responsibility pay little or no attention to politics and public policy. This omission suggests that they may regard government as irrelevant.

This perspective is shortsighted, considering the historical record. In the roughly 30 years that have elapsed since the major expansion of federal environmental regulation, the United States has made significant progress in improving environmental quality. Now what portion of this improvement can we attribute to better corporate environmental management? Or, to ask the same question differently, without the expansion of federal, and in many cases local, regulation as well during the 1970s and 1980s, how much improvement in environmental quality would have taken place?

Government regulation, for all its well-documented shortcomings, is responsible for virtually all the progress that we have made. Indeed, environmental regulation deserves to be considered one of the most important domestic policy successes in the postwar period. To take just one example, the average automobile emits roughly 97 percent less pollution per mile driven then it did in 1970. What role has corporate environmentalism played in this astonishing accomplishment, which has been critical in improving urban air quality? The answer is simple: none whatsoever. It is entirely due to the Clean Air Act Amendments of 1970, 1977, and 1990—legislation, it is worth recalling, enacted over the bitter opposition of the automobile industry.

Admittedly, corporate environmentalism is, by many accounts, a relatively recent phenomenon. It is possible that in the future, consumer, shareholder, public, and employee pressures may play a greater role in improving corporate environmental performance than they have in the past. For example, the production of hybrid vehicles by a few automotive manufacturers is not a response to public policy. But compared with government regulation, the role of voluntary corporate initiatives will remain modest. Can anyone seriously believe that American greenhouse gas emissions can be reduced—or even that their rate of increase can diminish significantly—without federal regulation that changes the incentives of corporations and consumers? Or, to take an even more urgent environmental problem, how can more "responsible" firms adequately address the problem of unsustainable fishing in international waters? Isn't an effective international agreement essential to address this classic "tragedy of the commons" problem?

If we want managers to become serious about creating more "sustainable" corporations, then we should challenge them to support, or at least not oppose, government regulations that will make it possible for them to behave more responsibly—without weakening their competitive position. The environmental payoffs of such regulation would be far greater than all the myriad "voluntary" or "enlightened" activities that fall under the rubric of corporate environmental responsibility.

Business lobbying is one of the most important ways corporations affect environmental quality. We need a broader definition of corporate responsibility that includes business political activity. I am aware of no measures of corporate environmental performance that incorporate the positions taken by firms on environmental policy issues. This is a serious shortcoming.

To be sure, some firms on occasion have supported more stringent environmental standards. But such cases are atypical. Needless to say, I do not expect firms to support public policies that make them less profitable, any more than I expect corporations to market greener products that reduce their market share or profit margins. But I think much contemporary writing on corporate environmentalism has done the cause of environmental protection a major disservice by ignoring the critical role of government. Government regulation may not be sufficient, and it certainly needs to be improved. But there is no effective substitute for it.

The business community itself benefits from this myopia. It is clearly in the interest of companies to persuade us that business is capable of significantly improving environmental quality on its own, because the alternative is more regulation. Their position makes sense. What is disturbing is that many influential writers outside the business community accept business claims so uncritically.

Reinhardt does an excellent job of critically evaluating the substantial and evidential body of literature on whether it pays to be green. He correctly notes the shortcomings of such studies. But his criticisms do not go far enough, for the question itself is largely irrelevant. An important and distinctive strength of Reinhardt's work is his claim that we should view corporate environmental decisions in the same light as any other business decisions. Let us then apply this reasoning to the subject of whether it pays to be green. Would it make sense to ask if marketing, research and development, or overseas expansion paid? The answer is obvious: it would not. The reason is a simple one: sometimes they pay and sometimes they do not. Or to be more precise, whether they pay depends on wide variety of other factors, some of which are outside the control of management and others of which are either unknown or unknowable. Why should we ask any or all corporate environmental programs to meet a criteria that we apply to virtually no other business activity? All business decisions are made in the face of uncertainty; nothing always pays. The critical question is always, when does a particular business decision pay?

Moreover, in many cases, the costs of environmental management are easier to measure than are their financial consequences. How, for example, do we assess the financial benefits of avoiding negative publicity, or of better employee morale? Nor does this make environmental management unique. The benefits of many things corporations do are difficult to measure. Can we calculate, for example, the impact of purchasing another corporate jet or a new corporate headquarters on the creation of shareholder value? Indeed, it has even proven challenging to measure the impact of CEO compensation on shareholder equity. Why should we expect environmental expenditures or programs to be any different?

A common way of measuring the financial benefits of environmental management is to assess the relative performance of "green" or "environmentally responsible" mutual funds or indexes. Not surprisingly, such measurements have produced mixed results: sometime such funds perform better than the market, and other times they do not. But there is no reason why we should expect them to consistently do so, as in the long run, hardly any managed funds outperform the market.

Moreover, if greener companies regularly did perform better, then every fund manager would weigh such stocks more heavily in his or her portfolio—in which case, paradoxically, there would no longer be any difference between the performance of "ethical" and "normal" funds. The same applies to management decisions: if it could be demonstrated persuasively that being green paid, then presumably every rational firm would become greener, in which case, of course, being green would no longer be a source of competitive advantage.

Instead of devoting so much academic research in a fruitless effort to demonstrate that being green pays, scholars would be better advised to focus more on studying a different question: namely, what explains improvements in corporate environmental performance? Answering such a question would enable us to tease out the relative importance of the wide variety of factors that drive corporate environmental decisions. These include, in addition to government regulation, pressures from consumers, investors, and employees; the fear of adverse publicity; the search for "goodwill"; and not least, management values, corporate culture, and social norms.

Were we to do so, I suspect we might find that many such decisions have little or nothing to do with the creation of shareholder value. One of the shortcomings of approaching environmental performance as a management problem is that it focuses too much of our attention on the search for competitive advantage. Yet many management practices have little or nothing to do with this admittedly central business objective.

Take, for example, corporate philanthropy, at least before it became "strategic." Large, publicly trading corporations in the United States historically have allocated roughly 1 percent of their pretax earnings to such expenditures. Yet it would be difficult to demonstrate that few if any of these expenditures have benefited shareholders. Nor were many intended to do so—except very indirectly. Yet corporations have continued to make philanthropic expenditures, and significant management time still is devoted to raising funds for the United Way.

Another example is the response of business to the "urban crisis." As a response to the ghetto riots of the late 1960s and early 1970s, many companies supported programs to address the exclusion of many inner-city racial minorities from the economic mainstream. Some of these efforts were effective, many were not, and most have long since been abandoned. But few if any such programs represented a source of competitive advantage. Nor did many managers at the time believe that they would. Yet they still committed resources to them. Why?

Notwithstanding competitive pressures, many corporations have slack resources. Such resources enable managers, if they choose, to establish policies and programs as a response to changing social norms. Managers engage in such activities because they have been persuaded, in part by peer pressure, that they are legitimate and appropriate. Some portions of corporate environmental policies and programs fall into this category. They can be regarded as the contemporary equivalent of the business response to the urban crisis. To ask if they "pay" may be to ask the wrong question. Sometimes corporations engage in activities because it is expected of them. Better understanding such norms and pressures—and their limitations—can provide us with a more complete picture of the dynamics of corporate environmentalism.

Finally, we need to place in perspective the increasing interest of many corporations in environmental responsibility. There is considerable evidence that, as a response to public pressures, many large American- and European-based multinational companies have improved their environmental performance in many developing countries. These efforts are important and commendable. But these initiatives do not, as many business school faculty and executives would have us believe, herald a new era of corporate responsibility.

On one important dimension, many firms have become less responsible: they have become much more willing to lay off workers than in the past. And those who retain employment often have faced a reduction in benefits, such as pensions and health care.

Whether, on balance, reducing pollution near an oil production site in Africa adequately compensates for reduced job security in Texas is an issue on which reasonable people can disagree. But we should not celebrate the significance of former and ignore the social costs of the latter.

Summary of Discussion on Corporate Social Responsibility and Business

This discussion focused on two questions: the role of ethics in corporate social responsibility (CSR) and the role of CSR in environmental policymaking.

Ethics and CSR

Paul Portney began the discussion by asking how should we think about the ethics of environmental protection in developing countries where more environmental protection would lead directly to less employment. Eric Orts responded that multinational firms increasingly adopt uniform environmental and worker safety standards wherever they operate. Although uniform standards may not be optimal economically, they can be justified on the deontological premise that all human lives should have equal weight. For Orts, this argument appeared to have both positive and normative significance. On the positive side, it purports to explain why firms engage in profit-sacrificing CSR; on the normative side, it offers an answer to the question, "should they?"

John Coates introduced the example of slavery in Sudan in support of the positive argument. In his view, firms choose not to operate in Sudan not because of reputational costs, but because managers think it is wrong. This way of thinking illustrates how moral reasoning plays an important role in CSR "across some range of highly predictable harms."

Max Bazerman reinforced this argument by referencing a body of psychology literature that shows that people intuitively identify some things as simply not tradable. Likewise, he cited an argument by Richard Thaler in the behavioral finance literature that holds that it can be welfare-enhancing for the state to engage in libertarian paternalism. Similarly, Bazerman suggested that it might make sense for firms to make decisions about appropriate levels of risk in situations where the trade-offs between risk and wage are not intuitive to many workers. Robert Stavins pointed out that the same reasoning commonly is used to justify government policy, as opposed to private-sector decisionmaking in the environmental realm.

The discussion then focused on whether people are better off working in a dirty plant or not working at all. What is the deontological case, asked Bruce Hay, for forcing workers to accept a risk reduction when they would prefer higher wages and more risk? If workers prefer higher wages and more risk, then a policy that compels them to accept less risk *and* lower wages would, in fact, reduce their welfare. Paul Portney added that individuals in different circumstances make different choices. Respect for those choices, he argued, has a strong ethical foundation. Daniel Esty pointed out that Portney's argument did not hold for environmental harms that have international consequences. For example, India should be free to set worker protection standards, but also should be expected to adhere to international guidelines regarding emissions of ozone-depleting chemicals.

Stavins tried to resolve the apparent contradiction between Coates' point about slavery in Sudan and the views expressed by Hay and Portney regarding recognition of workers' preferences. Slavery, he pointed out, is a special case precisely because slaves cannot express their preferences through labor markets. Such a situation is fundamentally different from cases in which firms and workers trade off wages against environmental quality. In the case of slavery, ethics clearly should constrain behavior. In the case of making trade-offs between wages and the environment, the preferences of market participants should hold sway.

Orts agreed with this, acknowledging that wages should vary across countries according to workers' preferences and local economic conditions. But what should not vary, he maintained, is the risk that workers are killed or injured. Do attitudes toward risk and death really differ across workers in developing and developed countries?

CSR and Environmental Policy

Coates shifted the discussion to the role of CSR in environmental policy by claiming that environmental policy has been formulated in one of two ways: either in response to a crisis or with the active support of the private sector. Thus CSR may play a central role in the formulation of environmental policy in the absence of crisis. Other discussants, however, noted several prominent examples of major U.S. environmental policies, such as the 1970 Clean Air Act, that were neither induced by crisis nor widely supported by firms. Coates' view, they suggested, left out the powerful role played by environmental advocacy groups, a point Coates conceded, although he noted that active involvement of the private sector has characterized environmental policy shifts in the last 20 years.

A related exchange focused on the historical role of CSR. David Vogel asked what explains the enormous gains in environmental protection over the past 35 years, and suggested that public policies plus technological change explain nearly all of the improvements, with CSR playing only a very small role. Esty disagreed, arguing that shifts in attitudes and behavior of corporations had paved the way for public policies.

Others maintained that the global scale of both firms and environmental problems means that the past may not be a good guide to the future. Because of governments' relative lack of power in international contexts, firms may have gained considerable leeway to trade off the environmental impacts of their activities against profits. To some, this suggested that businesses that see reasons for protecting the environment may play an increasingly important role in shaping societal responses to environmental challenges in the future.

Although the group reached no consensus about the magnitude and significance of trends toward voluntary environmental reforms, most agreed that business school students place enormous faith in the power of enterprise to solve social and environmental problems. Thus even if the empirical record offered little support for the hypothesis that voluntary action by firms is supplanting traditional environmental regulation, the notion that firms can and should "be part of the solution" is an important concept among future business leaders.

In this discussion of CSR and environmental policy, the idea surfaced again that firms may use voluntary measures to manipulate regulatory outcomes in two distinct ways. First, firms may use CSR strategically to create a regulatory environment that offers them some advantage over their competitors, as, for example, DuPont's efforts—cited by Forest Reinhardt—to change CFC usage prior to regulatory change. And second, firms simply may use CSR to check the emergence of regulations that would be costly to all firms, as with the example of support for "Responsible Care."

Index